Vera Axyonova, Florian Kohstall, Carola Richter (eds.)
Academics in Exile

The Academy in Exile Book Series | Volume 2

Editorial

The Academy in Exile Book Series is edited by Vanessa Agnew, Kader Konuk and Egemen Özbek.

Vera Axyonova is a Marie Curie REWIRE fellow at the University of Vienna and principal investigator of the project "Expert Knowledge in Times of Crisis". Previously, she worked in research, science management and policy consulting. She is co-founder of the ECPR Research Network on Statehood, Sovereignty and Conflict. Her research interests include expert knowledge production, international conflict management and norm transfer as well as the role of civil society actors in policy processes, focusing mainly on the post-Soviet space.

Florian Kohstall is the head of the "Global Responsibility" program at the Center for International Cooperation at Freie Universität Berlin. He is the founder and director of "Academics in Solidarity", a transnational program for displaced academics, and coordinator of Welcome@FUBerlin for refugee students. His research interests include higher education reform, especially in the Middle East and North Africa, varieties of internationalization and the role of academic freedom in discourses and practices.

Carola Richter is a professor for international communication at Freie Universität Berlin. In her research, she focuses on media systems and communication cultures in the MENA region (Middle East and North Africa), foreign news coverage, media and migration as well as on public diplomacy. She is the co-founder of AREACORE, the Arab-European Association of Media and Communication Researchers, and director of the Center for Media and Information Literacy (CeMIL) at Freie Universität Berlin.

Vera Axyonova, Florian Kohstall, Carola Richter (eds.)

Academics in Exile

Networks, Knowledge Exchange and New Forms of Internationalization

[transcript]

The publication of this volume has been underwritten by generous support from the German Federal Ministry of Education an Research (BMBF) and the Academics in Solidarity program.
This book is freely available in an open access edition thanks to funding by the Freie Universität Berlin.

Bibliographic information published by the Deutsche Nationalbibliothek
The Deutsche Nationalbibliothek lists this publication in the Deutsche National-bibliografie; detailed bibliographic data are available in the Internet at http://dnb.d-nb.de

First published in 2022 by transcript Verlag, Bielefeld
© **Vera Axyonova, Florian Kohstall, Carola Richter (eds.)**

Cover layout: Maria Arndt, Bielefeld
Cover illustration: ID 3796908, www.pixabay.com

Print-ISBN 978-3-8376-6089-0
PDF-ISBN 978-3-8394-6089-4
https://doi.org/10.14361/9783839460894
ISSN of series: 2701-8970
eISSN of series: 2747-318X

Contents

The emergence of a "third space"?

South-South perspectives

Acknowledgements

The editors and authors would like to thank the German Federal Ministry of Education and Research (BMBF) that has provided generous financial support for the preparation of this book as part of its funding of the Academics in Solidarity program. Moreover, Freie Universität Berlin kindly supported the Open Access Publication with its funds. Vera Axyonova's work on this book project was in part funded by the European Union's Horizon 2020 research and innovation program under the Marie-Skłodowska Curie grant agreement No. 847693.

Our sincere thanks also go to Leona Ley, Sabrina Schotten and Frida Siering for their tremendous help in formatting and copyediting the manuscript.

Academics in Exile
Networks, knowledge exchange, and new forms of internationalization in academia. An introduction

Vera Axyonova, Florian Kohstall & Carola Richter

Since the founding of the modern university, academic mobility has been a key element in knowledge transmission and production. The *Authentica Habita Act* or *Privilegium Scholasticum* of 1158, among others, guarantees the freedom of travel for scholars. It is among the founding documents concerning academic freedom. The pursuit of knowledge has always involved moving toward unknown places and sources. Yet, when academic mobility is placed in the context of exile, it becomes a paradox because exile is the opposite of free movement. It is the result of forced mobility and severe limitations on academic freedom in places that once may have been hubs of learning and knowledge production. When academics are in exile, we can expect that academic freedom is under severe pressure in their countries and their institutions of origin.

One of the primary objectives of this book is to unpack the paradoxes of exile and contrast the often romanticized pictures of émigré life with the daily experiences of academics who did not leave their home countries voluntarily. When we think about academics in exile, prominent figures such as Albert Einstein, Hannah Arendt, and Edward Said come to mind. While the literature often stylizes them as heroes of their time, most exiles have struggled and continue to struggle with border regimes, bureaucratic hurdles, and the burden of adapting to new social and academic environments. Moreover, many exiled academics remain *de facto* invisible unless they are mentioned in one of the support organization reports for at-risk and displaced scholars. What Edward Said (2000, p. 173) wrote in his famous reflections on exile years ago remains valid:

Exile is strangely compelling to think about but terrible to experience. It is the unhealable rift forced between a human being and a native place, between the self and its true home: its essential sadness can never be surmounted. And while it is true that literature and history contain heroic, romantic, glorious, even triumphant episodes in an exile's life, these are no more than efforts meant to overcome the crippling sorrow of estrangement.

These years in the early 2020s are a particularly apt time to inquire into the lived experiences of academics in exile. Never before have so many academics who had to leave their home countries found refuge in Germany. Some observers even now speak of Berlin as the capital of the Arab Exile (Ali, 2020). While countries such as the USA and Great Britain have longer histories of receiving threatened scholars, few comprehensive studies exist on the professional trajectories of these scholars and their contributions to knowledge production and exchange. We know little about the individual pathways and the networks of these scholars in their home and host countries. How do their forced migration experiences affect their professional careers? To what extent do different support programs allow them to resume their research activities, and thus produce and share new knowledge in host environments? How do these scholars navigate the often quite different academic systems and an academic labor market that is by far not as internationalized as it would seem? Do networks of exiled scholars have an impact on transnational scientific collaboration, and do their presence and activism contribute to the reform and internationalization of academia? It is time to address these questions in order to understand the bigger picture of how involuntary academic mobility impacts transnational knowledge exchange and how the emergence of exiles as new actors in internationalization processes affects higher education across the globe.

This book sheds light on these questions by combining two perspectives: that of displaced and at-risk scholars, examining their individual trajectories, lived experiences, professional networks, and contributions to knowledge production, and that of host country institutions, uncovering the opportunities and challenges of what has been termed "forced internationalization" (Ergin et al., 2019) experienced by university leadership, research funding organizations, and government agencies. Before discussing these two perspectives more in-depth and thus focusing on what happens *in* and *as a result of* exile, we must engage with the reasons for involuntary academic mobility – what pushes scholars *into* exile.

Why scholars flee: From "inner exile" to "physical exile"

Restrictions on academic freedom and the struggles that result from these restrictions are among the main reasons for scholars' displacement worldwide. These struggles are often long-term because the constraints to academic freedom can be manifold. According to Grimm & Saliba (2017, p. 48), academic freedom includes the right to expression, distribution of research results through publications and conferences, teaching, and conducting research on any topic. The authors operationalize the constraints to academic freedom along personal, legal, and economic dimensions (p. 56). Personal aspects play out at the individual level, from travel restrictions to imprisonment or murder. Legal aspects include censorship or limits to the freedom of association. Finally, economic aspects refer to the denial of funding or termination of employment as a preemptive or retaliatory measure. These restraints occur to various degrees and often accompany times of change and crisis in a country.

In recent years, we have seen many of these restrictions levied in the Middle East and North Africa (MENA) region, but also elsewhere. In Iraq, Syria, Yemen, and Libya, devastating wars have led to the destruction of university buildings and academic infrastructure to the extent that "universities were barely able to continue to function in a regular manner" (Saliba, 2018, p. 313). In addition, security forces or militias pose physical threats to scholars through intimidation, imprisonment, kidnapping, and even torture and murder. In other cases, such as in Turkey and Egypt, scholars and, more broadly, intellectuals have been pressured because their political opinions opposed the regimes' interests. Saliba (2018, p. 314) speaks of a "systematic crackdown on critical students and researchers" in these countries, resulting in arrests, expulsions, or disciplinary actions against academics.

The outlined restrictions on academic freedom and the effects on individual scholars can force them to give up their work, refrain from publishing, or even leave the country. Thus, flight can take the form of an *inner exile*, which involves ceasing certain research that is considered sensitive or includes political opinions. It can also result in *physical exile*, pushing scholars out of their home countries. Kettler (2011, p. 204) argued that "political exile is not a metaphor of estrangement, but a political condition." This implies the need for scholars to relocate due to external pressure. This has hardly anything in common with voluntary academic mobility when scholars relocate for their research, field work, or fellowships. Exile takes on a very different kind of "forced mobility" than the one Cantwell (2011, p. 430) described as "coerced movement from job to job resulting from differences in pay and opportunity and normative pressures in internationalization."

Several organizations have recently engaged in monitoring the state of academic freedom worldwide. The New York-based organization Scholars at Risk (SAR) provides systematic vetting for endangered scholars. Based on this vetting, it records infringements on academic freedom in its annual reports. In 2020, SAR reported 342 attacks on academic freedom worldwide, among them 124 killings, violence, disappearances, 96 imprisonments, 52 prosecution cases, 30 losses of positions, and 7 travel bans (Scholars at Risk, 2021). Based on these findings, we can reasonably surmise that the actual number of these types of incidents is even higher on a global scale.

The Academic Freedom Index (AFI), developed collaboratively by the Global Public Policy Institute (GPPi), the Friedrich-Alexander-Universität Erlangen-Nürnberg (FAU), the Scholars at Risk Network, and the V-Dem Institute, since 2020 has provided a more general overview of the situation and measures academic freedom country by country.[1] It relies on expert opinions and paints the world map in assorted colors according to the state of academic freedom. The AFI authors distinguish between attacks on individual and institutional academic freedom. According to the AFI map, academic freedom is being diminished in many countries of the world as well as in Europe and the Americas. One might still question whether the overall degree of academic freedom has decreased in comparison with the state of affairs in the interwar period of the twentieth century or the time of the Cold War. Nevertheless, it is important to underscore the volatility of the situation, the fragility of academic freedom, and the entanglement of historic trajectories.

A "transnational historicization" of watershed moments for forced migration

In the past ten years, the exile theme has gained new prominence in academic and practitioner discourses. Varied factors account for this newly discovered interest in the phenomenon, which had long remained largely unattended. Among these factors are the political and social repercussions of the Arab

---

1   See the presentation at https://www.gppi.net/2021/03/11/free-universities.

uprisings. These uprisings, which started in Tunisia in 2010, fueled new hopes for democratic openings and freedoms of thought and expression. In places such as Egypt and Tunisia, they created a window of opportunity to reform the university and engage in free scientific inquiry (Kohstall, 2015). However, the promises of democratization did not bear fruit. In Egypt, where authoritarian rule was quickly reestablished, academics, especially from the humanities and social sciences, still suffer under severe restrictions. In Syria, Yemen, and Libya, the uprisings turned into armed conflicts, forcing millions to flee, among them many students and scientists.

By the end of the 2010s, Turkey had become another large "sender" of academics in exile. Many were part of the Academics for Peace (BAK in Turkish) initiative. This initiative was founded in 2012 by scholars concerned with the government's unlawful treatment of Kurds in Turkey's eastern provinces. In January 2016, BAK initiated a petition exclaiming, "We will not be a party to this crime!" The petition referred to the heavy-handed repression of the Kurds by the Turkish government and was initially signed by 1,128 Turkish academics. Turkish president Recep Tayyip Erdoğan immediately slammed these academics as "so-called intellectuals," calling on their universities to punish them (Hürriyet Daily News, 2016). He argued that the petition aimed at destabilizing the Turkish state and jeopardized the security of its citizens. The signatories were criminalized: dozens were immediately imprisoned, and hundreds lost their jobs (Baser et al., 2017, p. 275).

The mass exodus of scholars from Turkey was a watershed moment in the annals of academic exile because it illustrated how singular events in one country could lead to a mass displacement of politically oppositional and otherwise threatened scholars. In recent history, similar watershed moments – although under different political circumstances – have occurred in Venezuela with the tightening of civic spaces by President Hugo Chávez from the 2000s to early 2010s, in Belarus following the suppression of mass demonstrations against the autocratic rule and re-election of Alexander Lukashenko in 2020, in Afghanistan after the reestablishment of the Taliban regime in 2021, and in Ukraine (and Russia) during the war in 2022.

Historically, such watershed moments occur intermittently, often with unforeseen outcomes, both for the scholars themselves and for academic knowledge production. Scholarly accounts of these outcomes differ in their assessments of what exile means for academics as individuals when they encounter new university cultures (Dauphinee, 2013; Vatansever, 2020), and as a collective of knowledge producers who – through their exile experiences – change that very cultures, creating new research trends and traditions (Rösch, 2014).

These different assessments also find their way into this volume. **Azade Seyhan**, for instance, illuminates the history of German scientists and intellectuals who fled from Nazi Germany to Turkey and contributed to a substantial refurbishment of the higher education system. The exiled scholars made a strong and lasting contribution to the establishment of secular Turkish universities and the translation of knowledge from one cultural context into another. Seyhan also recalls the role of lesser-known German emigrants to the southern US states at that time, demonstrating how they participated in the preservation of knowledge that was under attack by the Nazis by adapting themselves to a completely different cultural and political environment. Through her reinvestigation of the archive of exiled knowledge, she wonders if the displaced scholars of today might have a comparable lasting effect on the translation of knowledge in their host countries or if their voices will remain marginalized.

The path into personal marginalization and the limited possibilities for knowledge translation are illustrated in another contribution to this volume. **Rika Theo** and **Maggi Leung** focus on parts of the 1965 student generation in Indonesia, which was soon to become the country's new elite after studying abroad but could not return home because of the Suharto lead military coup that abolished the previous political regime. Theo and Leung vividly show the precarity, isolation, and degradation of professional prospects faced by these young academics, especially at the beginning of their lives in exile. Yet, the authors also demonstrate how – with time – these academics were able to establish networks with later generations of Indonesian students abroad and thus nourished a collective memory and passed on an alternative narrative of Indonesian politics that was not told in the official textbooks. The fate of these academics recalls other examples of students who became victims of sudden political change, such as the "generation Tahrir," which preserves abroad the memory of the Egyptian and the Arab uprisings.[2]

Seyhan's chapter on German scholars in Turkey and the US and Theo and Leung's discussion of the Indonesian 1965 student generation document the complexity of "refugee routes" (Agnew et al., 2020). While Turkey received

---

2   For an example, see the documentation of the Génération Tahrir project at http://www.generationtahrir.net

many threatened scholars in the past and continues to offer refuge to academics from Syria, it now also belongs to the countries that force their own citizens into exile. The experience of Indonesian scholarship holders is a historic precedent for many intellectuals who could never return home. Largely isolated in the early years of exile, they later offered "refuge" to the alternative history of political developments in Indonesia, sharing it with younger generations. Reattending this entanglement of "refugee routes" and the "transnational historicization," proposed by Vanessa Agnew (2020, p. 18) in the first volume of this book series, is necessary to tell the manifold stories of academics in exile. It requires revisiting the places of flight and destruction as much as the places of arrival and reconstruction and involves a "travail de memoire," as well as reworking the individual and institutional memory.

Forced internationalization: The relationship of the protection of academic freedom, humanitarian intervention, and the quest for excellence

For a long time, political developments on the southern shores of the Mediterranean seemed to affect mainly the countries and societies there. Europe observed the transformations with interest but without much enthusiasm and engagement. Only with the so-called "summer of migration" in 2015, when hundreds of thousands of refugees entered the European Union did EU Member States face the immediate need to react. For countries such as Germany, 2015 became a watershed moment. Suddenly, the country could prove its "Welcome Culture" (Streitwieser et al., 2017), which was concisely stated in Chancellor Angela Merkel's famous sentence, "Wir schaffen das" ("We can do it"). Numerous institutions set up initiatives to support the newly arrived. Universities soon took center stage, launching language courses, mentoring programs, and labor market counseling for displaced students and scholars.

Whereas during the 1930s thousands of scholars fled from the Nazi regime to the US, Turkey, and other countries, in the mid-2010s, Germany became one of the first countries to receive scholars under threat. In 2015, these scholars were mainly from Syria, but then also from other countries of the Middle East, and soon – in a historic irony – mainly from Turkey. In this context, a new narrative gained ascendancy among non-governmental organizations (NGOs), funding agencies, and state institutions that were involved in support of threatened scholars: a narrative that would embed initiatives for hosting exiled scholars into the logic of protecting academic freedom worldwide.

While university infrastructure destroyed by war and forced displacement were the initial reasons for constituting scholar rescue programs in the 2010s, the need to react to the systematic prosecution of academics in countries not affected by violent conflict soon became the various programs' common theme. This trend was reinforced through the perception that academic freedom was in decline worldwide, including in Europe and North America. In Hungary, for instance, the Central European University (CEU) had to relocate its study programs from Budapest to Vienna after the adoption of the so-called "lex CEU"[3] by the Hungarian parliament in 2017, which nearly drove the university out of the country completely. In the US, several states' legislative attempts to curtail tenure in public universities, which had been introduced as a tool against arbitrary firing and retribution over a century ago, raised concerns about the gradual deterioration of faculty members' rights (Worthen, 2021). Furthermore, the reputation of academic institutions and public trust in science were damaged under the presidency of Donald Trump, who on many occasions challenged scientific evidence, creating additional pressures on scholars as knowledge producers. Under these conditions, the need to protect academic freedom became the dominant narrative in support programs for at-risk scholars, complementing and partly replacing the initial appeals for humanitarian intervention in response to war and forced displacement.

Despite its history, and unlike the UK or the US, Germany did not have large-scale initiatives for supporting threatened scholars prior to 2015. One of the first and most prominent programs set up for this purpose in the country was the Philipp Schwartz Initiative (PSI) of the Alexander von Humboldt Foundation. It was named after the Austrian pathologist Philipp Schwartz, who lost his professorship in Frankfurt under the Nazi regime in 1933 and founded the *Notgemeinschaft deutscher Wissenschaftler im Ausland* (Emergency Society of German Scholars Abroad) in Switzerland. Inspired by this historical act of solidarity, the PSI provides funds to German universities for hosting at-risk and displaced researchers and serves as an example of national support

---

3   "Lex CEU" was a modification to Hungary's 2011 higher education law, which targeted the operation of foreign universities in Hungary. It ruled that such universities could only operate in Hungary if they did so in their country of origin as well.

programs established throughout Europe after the "summer of migration" in 2015.

These programs are complemented by subnational and transnational initiatives, such as the Leipzig-based *Chance for Science*, aimed at matching refugee scientists with their German peers, *Academics in Solidarity*, based at Freie Universität Berlin and actively engaged in the mentoring, network-building and further training of displaced scholars residing in Germany, Jordan, and Lebanon, or the *Academy in Exile*, established in 2017 as a residency program for endangered scholars in Essen and Berlin (for a comprehensive overview of the latter, see Konuk, 2020). European-level initiatives, such as the *InSPIREurope* project launched in 2019, aim at cross-country coordination between national and subnational rescue and support programs.

The establishment of different programs to support scholars under threat is part of the phenomenon that Ergin et al. (2019) have dubbed "forced internationalization." Interestingly, similar to the "voluntary internationalization" of higher education, forced internationalization is expected to produce positive effects on the host academic environment through brain gain and an increased diversity of scientific perspectives. Yet, in contrast to the former, forced internationalization is not a proactive and deliberate policy pursued by universities and ministries of education. It can be framed as a humanitarian reaction to the forced displacement of academics resulting from violent conflicts, restrictions to academic freedom, and the systematic persecution of scholars. At the same time, scholar rescue and support programs need to be situated in a wider context that includes at least two other drivers behind internationalization processes, namely, the ambition of universities and funding organizations to play an active role in science diplomacy and the continuous quest for excellence as the DNA of the international research system.

At their annual meetings, the different organizations in support of at-risk scholars oftentimes put the worldwide protection of academic freedom at the core of their keynote speeches. Also, in countries where academic freedom is a sensitive issue, other narratives, such as the protection of cultural heritage in the MENA region, are brought to the fore to justify intervention in support of displaced scholars. In both cases, the discourse embraces a certain missionary statement: the defense of a common good that necessitates diplomatic action in other countries. Simultaneously, excellence in research and the potential for success in pursuing scientific work remain concurring references in the narratives promoted by science diplomacy. Moreover, academic excellence and the potential for success have been two important selection criteria in support schemes for at-risk scholars, which results in moral dilemmas and contradictions in implementing these programs.

In our volume, **Isabella Löhr** embeds an analysis of these dilemmas and contradictions in a reflexive approach to migration. She demonstrates how global asymmetries in knowledge practices and possibilities to establish (seemingly universal) standards of scientific excellence translate into selection processes of scholar rescue programs, which end up prioritizing academics with better employability prospects in host countries, at times to the detriment of those who are more at risk. Thus, the strive for excellence – measured by the criteria dictated by the academic job market – clashes with the humanitarian logic of providing support to those who need it most, creating a moral dilemma for participants in the selection processes. Löhr further argues that such selection mechanisms are nothing new or specific to scholar support programs. They are rooted in the "contested politics of migration" and border governance, whose primary function is encouraging the mobility of certain groups (e.g., highly skilled professionals) while immobilizing others.

Echoing Löhr's argument, **Lizzy Anjel-van Dijk** and **Maggi Leung** unveil the contradictions between a highly neo-liberalized (and largely commodified) higher education system and the support of exiled scholars in the Netherlands. They argue that the competition-driven academic system takes a high toll on researchers in general, but a much greater one on scholars who were forced to flee their home countries. The latter must adjust to the requirements of the Dutch academic job market within a short period of time in order to be employable in research and teaching, regardless of their personal circumstances. The different administrative and structural hurdles exiled scholars are facing suggest that their international experiences and alternative theoretical and methodological perspectives have little "value" in the Dutch research system, despite the official rhetoric behind the state-aided support programs.

When taking a closer look at the situation in Germany, the experiences of exiled scholars supported through the various programs also contrast with the dominant narrative of the "Welcome Culture." Despite the humanitarian credo of opening doors to displaced scholars, rescue and support initiatives cannot offer them long-term perspectives in a system largely based on short-term employment. In her chapter, **Aslı Vatansever** argues that in the highly competitive and polarized German higher education market, "exiled scholars

. . . are effectively incorporated into the reserve army of precarious academic workforce." This raises the question of to what extent the principles of academic freedom can really be upheld in light of high job insecurity. At least for Vatansever, the relatively easy entry of displaced scholars into the system through at-risk scholarship remains a lure (see also Vatansever, 2020).

The emergence of "third spaces" as a result of forced internationalization

While the arrival of displaced scholars in the German and European systems remains in flux, and it is too early for a final assessment, we argue that the processes of forced internationalization can have lasting effects on the academic systems in the host and home countries of exiled scholars as well as on the lives and intellectual legacies of the scholars themselves. Despite the obstacles faced by displaced researchers, they establish new academic networks and intervene in the complex fabric of knowledge production in their host environments. Although the politics of forced internationalization are reactive rather than proactive, the resulting opportunity structures can lead to intensified exchanges among epistemic communities that otherwise have few chances to intersect. The long-term effects thereof could go as far as bridging the global gap in knowledge production. By joining host academic institutions in the Global North, exiled scholars – many of whom are from countries of the Global South – bring with them ideas, concepts, and theories that are not necessarily part of the mainstream academic debates in the countries considered to be "core" producers of science (Demeter, 2020). Thus, these ideas, concepts, and theories enter the "core spaces," potentially altering the mainstream scientific discourses and complementing them with alternative perspectives and methodological approaches.

Furthermore, the exchange of knowledge facilitated through the processes of forced internationalization may question and shift the common understanding of what is "important," "excellent," and "useful" science. What may appear to be of crucial importance to scientists in the Global North might be of less relevance in other world regions (Kraemer-Mbula et al., 2020). And yet, rooted in colonial legacies of modern science and what has become known as "intellectual imperialism" (Alatas, 2000), theoretical approaches and concepts originating in the Global North dominate international scientific and policy debates. Enabling greater interconnections among knowledge producers across the invisible Global North–South divide, research internationalization – including that which results from involuntary academic mobility – can serve to empower those whose voices are less heard in the World Scientific System (Beigel, 2014). While this process will not reshape "the geography of science" (Skupien & Rüffin, 2020), it may still contribute to reducing the existing global inequalities in knowledge production.

Again, these developments imply a great deal of hardship for the individuals involved, and it may take time until they come to the forefront. The three chapters combined in this volume's section on the emergence of "third spaces" explore how scholars in exile struggle with pursuing their research under precarious conditions, how they establish new formal and informal networks in this environment, and how they create new spaces for teaching and research outside the formal arena of the university. Thus, they contribute to the goal of exploring new research foci and teaching formats and intensifying cross-regional knowledge exchange. Of course, these contributions also emphasize the hurdles that must still be overcome in order to facilitate the reentry into the academic market for displaced scholars, such as formal restrictions on teaching, dependence on "host professors," and language obstacles as well as the logic of the market itself.

In her contribution, **Ergün Özgür** emphasizes the challenges that displaced scholars encounter in Germany when trying to reenter the academic labor market, but also the opportunities that arise by their inclusion into German academia. Among the challenges are not only short-term funding opportunities and precarious working conditions, but also the question of language proficiency, healthcare, social security, and bureaucratic mechanisms. While most scholars get support from their host institutions, Özgür highlights that this support can vary from one institution to the other and in terms of how much experience they have with hosting scholars. In turn, hosting exiled scholars also has an effect on the host institution: it prompts universities and research institutes to develop more cooperative mechanisms and create additional opportunities for the internationalization of the research that otherwise would not have emerged.

Carola Richter, on the other hand, provides important insights into the career trajectories of at-risk scholars and their modes of communication in German exile. Her analysis has revealed major ruptures and changes in the networks of the investigated scholars due to their forced mobility. The chapter underlines the strong dependencies on host persons, which often contribute to a feeling of marginalization in the host country and the severing of relations with the home country. However, other scholars in the sample successfully strengthened their position in their host environments by relying on previously built transnational and newly formed connections. This seems especially true for the well-connected Turkish academics who today dominate the debate on the future of scholar rescue programs.

In addition to the existing state-sponsored programs, **Asli Telli** shows us how exiled scholars establish their own initiatives, cultivate and transmit alternative narratives, and revive otherwise "lost knowledge." Being one of the founders of such an initiative herself, she discusses the Mapping Funds Project, which acts as a repository for different national and transnational initiatives. She also maps the supportive network and relationships that exist among host and sending country institutions, research funding organizations, government agencies, and public/private donors. The chapter illustrates how stipends, which still remain the major source for integrating scholars into a new environment, are supplemented by grassroots initiatives, opening up new spaces for engagement on various levels in the host country but also in the home country. This provides hope that while addressing the needs of displaced scholars, these initiatives can address the issue of academic freedom at large, and as a result, create "widespread micro-organizations against a huge and multi-centered systemic power" (see p. 198) that restricts academic freedom.

Examples such as Off-University, Academy in Exile, and the New University in Exile Consortium further illustrate that parallel sites of knowledge production and transmission might emerge, producing courses and writings that are not part of the traditional curricula in the home or in the host countries of displaced scholars. This also tackles the question of how displaced and at-risk scholars might create "third spaces" in the academic system they become part of while in exile. These spaces are carved out by scholars negotiating their personal and professional sense of belonging in a new research environment. Similar to scholars who moved from the Global South voluntarily, displaced scholars have to establish themselves in Western academia (Martin & Dandekar, 2021), and additionally confront the reality of being driven out of their country into forced mobility. While exile often means a permanent and difficult to tolerate state of being in-between home and host country, between past and present, the third space may signal the obtaining of a future new home, at least in the sense that it opens new venues for participation and intervention.

Transnational perspectives on academic freedom and exile: The Global South as home and host

So far, we have touched upon different conceptions of academic freedom and the asymmetries in knowledge production between the Global North and the Global South. However, the forced migration of scholars does not occur solely along the North–South axis. On the contrary, many countries of the Global South become destinations of fleeing scholars. While the debate on academic freedom and support for at-risk scholars prevails in the Global North, it must be acknowledged that many countries where academic freedom is under threat are among the receivers and hosts of displaced scholars. The example of Turkey is one of the most illustrative in the recent history of exile. It is crucial to examine how countries and academic systems beyond the Global North respond to the displacement of scholars and to what extent Western conceptions of protecting academic freedom and humanitarian actions aimed to "save" scholars are echoed, emulated, and sometimes instrumentalized in these countries.

While the majority of contributions in this book concentrate on the German experience, we aim to open up the debate to the situation of displaced academics in other countries and regions. The goal is to move beyond a Eurocentric perspective emphasizing academic freedom and academic solidarity and to explore how the protracted displacement of scholars transforms academia in countries that find themselves among both senders and recipients of at-risk scholars. Through this transnational perspective, we intend to show that the boundaries between such categories as "home" and "host" countries are often blurred.

Thus, in this volume, **Olga Hünler** provides a historic account of the frequent dismissal of academics in Turkey during the twentieth century. Recurring military coups and coup attempts prepared the ground for the Turkish state's infringements on academics. Oftentimes, these infringements are linked to the rhetoric of reform and the construction of a national higher education system, which leaves little room for those who are regarded as "spoilers." Thus, the rhetoric on national emancipation through higher education and the interference in the autonomy of the university and the academic freedom of its staff are oftentimes closely intertwined. But while critical academics are dismissed, refugee academics, especially from Syria, are used to fill the staffing gap on the pretext of safeguarding the cultural heritage in the Middle East. Thus, Turkey also practices the politics of forced internationalization, although with its own set of motivations. Hünler further highlights how the state's reaction to the coup attempt of 2016 has produced positive side effects with respect to the emergence of "third spaces." Groups of Turkish academics, mostly purged from their posts with statuary decrees in the aftermath of the coup attempt, have launched independent initiatives to engage in research and teaching, creating new forms and sites for academic knowledge production.

Widening the geography to other countries in West Asia, **Nahed Ghazzoul** compares experiences of displaced Syrian academics in Jordan, Lebanon, and Turkey and examines the role of the host communities and international agencies in the articulation of these scholars' needs. In all three countries, the scholars are facing hardships in finding teaching or research opportunities to continue their academic work and sustain their livelihoods. This has halted their potentialities and deprived them of opportunities to advance their qualifications and competences. Ghazzoul argues that studying the circumstances of displaced Syrian academics more systematically would raise awareness in the international community about the challenges they encounter and thus enable a more apt response. She further suggests that such studies could engage Syrian academics as researchers. This would contribute to the improvement of their potential and the internationalization of academia in the host countries.

Finally, **David Gómez Gamboa** and **Lizzy Anjel-van Dijk** concentrate on scholars from Venezuela who left the country during the period of 2010–2020 as the result of the (re-)autocratization of the political regime and the exacerbation of the economic crisis and humanitarian situation. Most Venezuelan scholars migrated within the region, mainly to Colombia, Ecuador, and Chile. The chapter discusses the challenges and opportunities for individual scholars living abroad as well as for academia in the home and the host countries, resulting from this involuntary migration. The host countries offer the Venezuelan scholars better work and life conditions and, in turn, benefit from receiving a large number of highly qualified scholars who are in demand in these countries' rapidly developing higher education sectors. The states thus pursue, more or less, an open politics of (forced) internationalization by creating relatively attractive conditions for Venezuelan scholars to enter their academic systems. These politics are, however, driven by the host states' own demands rather than by the need for a humanitarian response. The negative consequences of these processes remain with Venezuela. On top of humanitarian and economic crises, the brain drain only adds to the uncertainty of the country's future.

The latter three contributions provide an important glimpse into the manifold facets of academic systems and academics under pressure where forced mobility has become the rule rather than the exception. They also attempt to draw our attention away from the spotlights of infringements on academic freedom toward higher education systems under permanent pressure and in permanent exchange. It is widely recognized that the impact of flight from Syria has been more disruptive for its direct neighbors Lebanon, Jordan, and Turkey, and it is still interesting to observe that researchers' mobility has barely been used to open up academic systems there, while Venezuela's neighboring countries offer relatively attractive conditions for incoming scholars.

Reflecting on exile with exiles and debating the praxis of receiving scholars with practitioners

Assuming a reflexive and participatory approach, this volume brings together authors with different academic backgrounds and mobility experiences. About half of the contributors have lived in exile themselves. Others have assumed lead positions in the management of support programs or have been involved in mentoring displaced and threatened scholars. What they all have in common is a critical reflection on a range of categories such as exile and scholar displacement, voluntary and forced mobility, academic freedom, knowledge production, and internationalization of research and higher education systems. They share the common vision that it is crucial to tell the experiences of academics in exile and make sense of the implications of these academics' re-arrival in the higher education sector.

While the primary goal of this book is to engage in the academic debate on exile, we hope that it will also provide food for thought for representatives of higher education institutions, funding agencies, and governmental entities involved in the practitioners' debates on academic freedom and the protection of scholars worldwide. A look at the Academic Freedom Index map indicates that more forced mobility of scholars is forthcoming. It is crucial to reflect on effective ways of inclusion for those on the move. Mobility and visiting programs have to integrate the specificities of forced migration, and higher education institutions need to put in place coordinated efforts to facilitate the arrival of displaced scholars. Some readers of this volume might not expect critical contributions from scholars who have benefited from rescue scholarships, but they will find that these scholars have made significant contributions. They will also be reminded that exile has often turned apolitical writers into political writers.

References

Agnew, V. (2020). Refugee Routes Connecting the Displaced and the Emplaced. In Agnew, V., Konuk, K., & Newman, J. O. (Eds.), *Refugee Routes: Telling, Looking, Protesting, Redressing* (pp. 17-32). Bielefeld: transcript.

Agnew, V., Konuk, K., & Newman, J. O. (Eds.) (2020). *Refugee Routes: Telling, Looking, Protesting, Redressing*. Bielefeld: transcript.

Alatas, S. H. (2000). Intellectual Imperialism: Definition, Traits, and Problems. *Southeast Asian Journal of Social Science*, 28(1), 23–45. http://www.jstor.org/stable/24492998

Ali, A. (2020). On the Need to Shape the Arab Exile Body in Berlin. *Dis:orient*, December 5. Retrieved from https://www.disorient.de/magazin/need-shape-arab-exile-body-berlin

Baser, B., Akgönül, S., & Öztürk, A. E. (2017). "Academics for Peace" in Turkey: A case of criminalising dissent and critical thought via counterterrorism policy. *Critical Studies on Terrorism*, 10(2), 274–96, https://doi.org/10.1080/17539153.2017.1326559

Beigel, F. (2014). Introduction: Current tensions and trends in the World Scientific System. *Current Sociology*, 62(5), 617-625. https://doi:10.1177/0013921145548640

Cantwell, B. (2011). Transnational mobility and international academic employment: Gatekeeping in an academic competition arena. *Minerva*, 49(4), 424–55.

Dauphinee, E. (2013). *The Politics of Exile*. Milton Park & New York: Routledge.

Demeter, M. (2020). *Academic Knowledge Production and the Global South*. Cham: Palgrave Macmillan.

Ergin, H., de Wit, H., & Leask, B. (2019). Forced Internationalization of Higher Education: An Emerging Phenomenon. *International Higher Education*, (97), 9-10. https://doi.org/10.6017/ihe.2019.97.10939

Grimm, J., & Saliba, I. (2017). Free research in fearful times: Conceptualizing a global index to monitor academic freedom. *Interdisciplinary Political*

Studies, 3(1), 41–75, http://siba-ese.unisalento.it/index.php/idps/article/vi
ew/17312

Hürriyet Daily News (2016, January 12). Erdoğan slams academics over petition, invites Chomsky to Turkey. Retrieved from https://www.hurriy etdailynews.com/erdogan-slams-academics-over-petition-invites-chom sky-to-turkey-93760

Kettler, D. (2011). A paradigm for the study of political exile: The case of intellectuals. In Stella, M., Štrbáňová, S., & Kostlán, A. (Eds.), *Conference Proceedings: Scholars in Exile and Dictatorships of the 20th Century, Prague, Czech Republic, 24–26 May* (pp. 204-217). Prague: Centre for the History of Sciences and Humanities of the Institute for Contemporary History of the ASCR.

Kohstall, F. (2015). From Reform to Resistance: Universities and Student Mobilisation in Egypt and Morocco before and after the Arab Uprisings. *British Journal of Middle Eastern Studies*, 42(2), 59-73, https://www.tandfonli ne.com/doi/full/10.1080/13530194.2015.973183

Konuk, K. (2020). Academy in Exile. Knowledge at Risk. In Agnew, V., Konuk, K., & Newman, J. O. (Eds.), *Refugee Routes: Telling, Looking, Protesting, Redressing* (pp. 269-284). Bielefeld: transcript.

Kraemer-Mbula, E., Tijssen, R., Wallace, M. L., & McLean, R. (2020). *Transforming Research Excellence: New Ideas from the Global South*. Cape Town: African Minds.

Martin, S., & Dandekar, D. (Eds.) (2021). *Global South Scholars in the Western Academy: Harnessing Unique Experiences, Knowledges, and Positionality in the Third Space*. New York: Routledge.

Rösch, F. (Ed.) (2014). *Émigré Scholars and the Genesis of International Relations. A European Discipline in America?* London: Palgrave Macmillan.

Said, E. (2000). *Reflections on Exile and Other Essays*. Cambridge: Harvard University Press.

Saliba, I. (2018). Academic freedom in the MENA region: Universities under siege. In European Institute of the Mediterranean (IEMed) (Ed.), *IEMed. Mediterranean Yearbook 2018* (pp. 313-316). Barcelona: IEMed. https://www. iemed.org/observatori/arees-danalisi/arxius-adjunts/anuari/med.2018/A cademic_Freedom_MENA_Ilyas_Saliba_Medyearbook2018.pdf

Scholars at Risk (2020). *Free to Think: Report of the Scholars at Risk Academic Freedom Monitoring Project*. New York. Retrieved from https://www.schol arsatrisk.org/

Skupien, S., & Rüffin, N. (2020). The Geography of Research Funding: Semantics and Beyond. *Journal of Studies in International Education, 24*(1), 24-38. doi:10.1177/1028315319889896

Streitwieser, B., Brueck, L., Moody, R., & Taylor, M. (2017). The Potential and Reality of New Refugees Entering German Higher Education: The Case of Berlin Institutions. *European Education, 49*(4), 231-252, DOI: 10.1080/10564934.2017.1344864

Vatansever, A. (2020). *At the Margins of Academia. Exile, Precariousness, and Subjectivity.* Leiden: Brill.

Worthen, M. (2021, September 20). The Fight Over Tenure Is Not Really About Tenure. *The New York Times*. Retrieved from https://www.nytimes.com/20 21/09/20/opinion/tenure-college-university.html

Histories of knowledge translation

Exile in a translational mode
Safeguarding German scholarship in Turkey and the United States during the Nazi reign

Azade Seyhan

> To write this German book, I had to emigrate in 1933 from a Germany whose stifling atmosphere after Hitler's conquest left me no air to breathe. The most important and pressing task imposed by the catastrophic world situation upon historian and sociologist alike, it seemed to me, was to determine just what had really happened and just what position we really occupy in the historical continuum.
> Alexander Rüstow (1980, xxiii)

Alexander Rüstow, a classicist by training and a Socialist by calling, and one of the key political figures of the Weimar Republic, narrowly escaped to Istanbul, when his efforts to stop Adolf Hitler from seizing power in 1933 failed and his family home in Berlin was ransacked by the Gestapo. He was almost 50 at the time, and during his 16-year exile in Turkey, he joined nearly 200 German-Jewish and antifascist academics and intellectuals, who were offered positions at the University of Istanbul and the University of Ankara. They taught in several fields, and their multiple areas of expertise embodied the epitome of the German ideal of *Bildung*,[1] a training that is almost impossible to receive today. As the above quote from the foreword to the abbreviated English translation of his multidisciplinary magnum opus *Ortsbestimmung der Gegenwart. Eine universalgeschichtliche Kulturkritik* (Situating the Present. A Universal Historical Critique of Civilization, 1950, 1952, 1957) indicates,

---

1   I am referring here to the concept of *Humboldtisches Bildungideal*, literally, the Humboldtian educational ideal, that emerged in the early nineteenth century and saw as its goal the achievement of both comprehensive general learning (including Latin and ancient Greek) and cultural knowledge.

Rüstow saw his enforced exile as a necessary condition for the comprehension and critique of the temporal forces that determine our situatedness in history. The overarching question of this momentous work was: How did we arrive at this catastrophic moment in time?

The first part of my inquiry concerns the historical conditions that necessitated the exodus of German culture to an unlikely destination. When Hitler seized power (*Machtergreifung*) in 1933 in a Germany besieged by political and financial crises, he swiftly moved to expel all scholars of Jewish heritage from their posts and effectively suspended every form of autonomy at the university. Among those dismissed from their jobs was the Hungarian-born Frankfurt pathologist, Dr. Philipp Schwartz, who fled with his family to Switzerland. In March 1933, Schwartz established the Zürich-based *Notgemeinschaft deutscher Wissenschaftler im Ausland* (Emergency Assistance Organization for German Scientists Abroad) to help Jewish and other persecuted German scholars secure employment in countries prepared to receive German refugees. The main focus of this chapter is the unique encounter in various translational modalities between academics exiled from the Third Reich and the Turkish institutions of higher learning that offered them refuge.[2] While a major part of my inquiry centers on the relatively unknown work of the German intellectual exiles in Turkey, I selectively compare their experience with that of the German scholars who immigrated to the US, not so much to stress similarities or differences, but rather to establish a heuristic premise that illustrates the different paths that intellectual transport and translation take. However, the fortunes of a virtually forgotten group of German scholars, who did not have the connections and clout of such well-known figures as Theodor Adorno, Hannah Arendt, or Herbert Marcuse, resemble those of the German émigré academics in Turkey. Like their compatriot exiles in Turkey, who landed in uncharted cultural territory, this group found refuge and employment in the "other" US, that is, the American South, which is not only geographically separated from the North by the Mason-Dixon line but was and, to some extent, still is also culturally, socially, and politically positioned against it. The story of these

---

2   For a comprehensive historical and critical account of the German academic exile to Turkey from Nazi Germany, see the memoirs of the exiles, such as Neumark (1980) and Hirsch (1982). An early academic study by Horst Widmann (1973) provides a full list of German-speaking academic exiles at the universities of Istanbul and Ankara with short biographies and comprehensive bibliographic sources.

émigré scholars, who took positions in traditionally Black colleges, needs to be remembered in the larger history of academic exiles. Therefore, I add an excursus on these scholars, who went "From the Swastika to Jim Crow."

The virtually unstoppable phenomenon of refugee flow to Western Europe and particularly to Germany bears witness to a historical irony, for during most of the last century Europe itself was a site of exodus, embroiled as it was in the inferno of two world wars. Of all European countries, Germany arguably represents the most radical transformation of a geography of emigration into one of immigration within the span of a few decades. Since most current academic research has concentrated predominantly on the immigration to Germany from the Middle East and the countries of the Mediterranean basin and on the challenges of non-integration and faltering acculturation, the archive of exile histories in and out of Germany remains incongruous and often incomplete. A more complete history would interconnect communities of research to negotiate the limitations as well as the inclusiveness of the archive. To that end, I examine the impact of German émigré scholars on Turkish and American academia and politics through a limited comparative assessment that highlights the correspondences between forms of exilic scholarship on opposite cultural shores as well as the different conditions in which exiled academics sought to preserve an intellectual legacy under threat.

While the critical gains of postcolonial theory and discourses on identity politics afford valuable insights into the unprecedented scale of migratory movements of our time, the historical contexts of displacement and deportation cannot only or consistently be abstracted from late twentieth and early twenty-first century narratives of exile. It behooves the critic and the historian to expand the domain of exile studies not only geographically but also historically, where acts of transport and translation across borders and epochs offer critical insight into the current large-scale population displacements. While languages and cultures in multidirectional movements destabilize paradigmatic unities, this instability enables the entry of once obscured texts into the archive, thus contributing to a new conceptualization of the transnational.

Redressing the archive of exile scholarship

Although the exodus of German scholars and writers to other European countries and particularly to the US has been extensively studied, the long-term sojourn of many noted academics and artists in Turkey has received scant critical attention. This situation changed somewhat when exile scholarship "discovered" that Erich Auerbach, who in 1935 was dismissed from his chair in Romance Philology at the University of Marburg, had written his magnum opus, *Mimesis. Dargestellte Wirklichkeit in der abendländischen Literatur* (1946) during the 12 years of exile he spent in Istanbul. The 50th anniversary edition of the English translation, *Mimesis: The Representation of Reality in Western Literature* (Auerbach, 1953), was issued in 2003 with an introduction by Edward W. Said.[3] Its publication generated some interest in and curiosity about other exiled German professors in Istanbul. There are notable publications, even documentaries, on individual figures such as Rüstow, Ernst Reuter, who before his exile was the Socialist mayor of Marburg and after his return to Germany, twice the mayor of Berlin, and the iconic Berlin architect Bruno Taut. However, there is as yet no critical study of the collective contribution of exiled professors to the sociocultural fabric of the host country. Among the other German academic exiles in Turkey during the 12 years of what was supposed to be a Thousand Year Reich, were the Romanist Leo Spitzer; philosopher Hans Reichenbach; Fritz Neumark, a prominent economist who taught at the University of Istanbul and served twice as *rektor* of the University of Frankfurt upon his return to Germany; Ernst E. Hirsch, professor of commercial law and a widely published legal expert, who was elected *rektor* of the Berlin Free University after his return to Germany; Georg Rohde, classical philologist, who played an important advisory role in the *Dünya Edebiyatından Tercümeler* (Translations from World Literature) series, inaugurated by the Turkish Minister of Education Hasan Âli Yücel[4]; the renowned sculptor Rudolf Belling, who had been fired from

---

3   This edition includes a long introductory essay by Edward W. Said, as well as an essay, translated into English for the first time, by Auerbach in response to his critics.

4   Yücel was the Turkish Minister of Education from December 1938 to August 1946. A linguist, philosopher, educator, and parliamentarian, he was considered a leading Turkish humanist. He is credited with establishing the *Köy Enstitüleri* (Village Institutes), which enabled the village youth to train as elementary school teachers close to home. Since Yücel oversaw the "Translations from World Literature" book series, which issued translations of both Western and Eastern classics, students at the Village Institutes were required to read 25 classical novels a year. The curriculum of the Village Institutes combined traditional educational subjects, including music instruction, with practical courses, such as farming, husbandry, carpentry, and home economics. Many noted writers and intellectuals were educated at these institutes,

his position at the Berlin Academy of Fine Arts for being a representative of *entartete Kunst* (degenerate art) and then appointed by Atatürk himself as chair of the sculpture department of the Istanbul Academy of Fine Arts; Rudolf Nissen, formerly professor of surgery at the University of Berlin, who headed the surgery department of the Medical School of the University of Istanbul from 1933 to 1939 and trained numerous Turkish professors and physicians; Paul Hindemith, musician and composer, who helped establish the Ankara State Conservatory; Carl Ebert, theatrical producer and director, who founded and directed the Ankara State Opera Company; and Eduard Zuckmeyer, a legendary music pedagogue. The list goes on and contains many other names, mostly scientists and medical doctors. This chapter of German intellectual history and its role in instituting a prescient transcultural and translational field of knowledge still awaits critical remembrance.

The dismissal of German-Jewish and antifascist professors from their posts coincided with the radical reform movements Kemal Atatürk (1881–1938), the founder of the modern Turkish republic and its first president, had undertaken in an ambitious modernization project, which included a top-to-bottom university reform. Along with Rudolf Nissen and Pedagogy Professor Albert Malche of Geneva, Schwartz visited Turkey in July 1933 and convinced the young Minister of Education, Reşit Galip, that the distinguished refugee professors would be instrumental to the success of the Turkish university reform. The visiting committee compiled a list of names for Galip, who persuaded Atatürk to personally support the project. As a result, a legion of anti-Nazi German, German-Jewish, and Austrian-Jewish scholars, artists, librarians, and teachers left Germany to accept various positions in the education sector in Turkey. Since Germany would have been reluctant to allow a massive exodus of scholars, it was decided that the contracts would be signed in a neutral country, in this case, Switzerland. The refugee professors were given long-term (up to five years) renewable contracts, which stipulated that they learn Turkish within three years and lecture in Turkish. However, in most cases, these requirements were not enforced.

which were closed in 1954 due to political pressure on the grounds that they were perpetuating left-wing ideas.

Conceptual premise

Historical conditions that necessitate the preservation of intellectual heritages through transport and translation contribute to a renewed understanding of texts that shape cultural movements across borders. In exile, the wandering culture is subject to translation in many senses of the term. Starting from this premise, I analyze exile scholarship in and as translation; the role of translation in the economies of a national culture; the imperative of cross-disciplinary work in exile; and the conditions for the emergence of an alternative critique of modernity at a non-European site. An investigation of how linguistic and cultural disparities between the home and the host lands can be negotiated may yield a model of successful integration of exiles and refugees in their respective lands of immigration.

What is of significance about the German academic exodus to Turkey is that the émigré professors were not considered displaced persons but reformers, and their role in the transformation of the university went beyond academic walls, leaving an enduring legacy in Turkish sociocultural life. Although the possibility of a humanist practice of *Bildung* was foreclosed by the Nazi takeover in Germany, this intellectual inheritance was kept in trust and safeguarded in translation both in Turkey and the US. The German and German-Jewish scholars, especially those associated with the Frankfurt School of Critical Theory, who found a safe intellectual harbor in the US, transported and translated a long legacy of German philosophical erudition into American academia. The careers of a considerable number of Frankfurt School members, such as the philosophers Theodor Adorno and Max Horkheimer; sociologist, psychoanalyst, and humanist philosopher Erich Fromm; philosopher Herbert Marcuse; sociologist and philosopher Leo Löwenthal; and journalist, sociologist, cultural critic, and film theorist Siegfried Kracauer, who is sometimes associated with the Frankfurt School, were spared thanks to the American institutions that welcomed them. At the end of the war, from the Frankfurt School in American exile, only Adorno and Horkheimer returned to Germany. Fromm spent his career at American universities and institutions and also taught at the National Autonomous University of Mexico. Only in the final years of his life did he return to Switzerland. I reiterate that I discuss the work of the exiled German scholars in the US only occasionally for purposes of establishing a broader critical framework, since there is already a substantial archive of scholarship on the German academic and artistic émigrés, who permanently or temporarily settled in the US.

Cultural exodus and the experience of the foreign

I contend that as opposed to contemporary and predominantly self-exiled scholars in the US, whose work reflects on their position as exiled subjects and engages in exilic politics, the German academics exiled from Nazi Germany are remembered for their lasting contributions to the host countries' educational systems and sociopolitical life. Since an analysis of cultural and symbolic values of translational acts requires a longitudinal range, I stress the enduring impact of those exiles, who spent all or most of their lives in the host country and integrated their cultural selves into the host institutions, while simultaneously reforming these. Although the number of natural scientists and professors of medicine in Turkish exile exceeded that of the humanists and their contributions to scientific research were considerable, the most visible legacy of the German academic émigrés in Turkey remains the establishment of a culture of translation in the broadest sense and the display of modern architecture as an allegory of the young nation. Therefore, I hope to foreground, in addition to Rüstow's work, the lesser studied contributions of Georg Rohde, the philosopher and psychologist Ernst von Aster, Ernst Reuter, and Bruno Taut. The last two city builders may not have name recognition beyond German and Turkish borders but are memorialized in the edifices they built in their Turkish exile. In the US, Hannah Arendt, Herbert Marcuse, Leo Löwenthal, and Siegfried Kracauer remained in their host country until the end of their lives and left an enduring impact on the cultural map of their respective fields. Before I discuss the lasting contributions of the émigré scholars to the sociocultural fabric of Turkey and the US, I would like to open a short parenthesis to describe the different routes the academic émigrés took to reach safe harbors. Such historical detail may or may not be relevant for displaced scholars of today but could prove useful as a precedent in assisting academics at risk.

Paths of exile

The realization of cultural transfer took two distinctly different paths to Turkey and the US. The exiled academics arrived in the US through circuitous routes and with the help of refugee agencies. Their journey was mediated by several different organizations, and the individual scholars and artists sought their fortune at different gates. In a great number of cases, their journeys were stalled at various stations of transit and their careers disrupted by multiple migrations. Hannah Arendt, for example, fled first to Czechoslovakia and then to Switzerland before settling in Paris. When France was invaded by the Germans, Arendt, her husband Heinrich Blücher, and her mother were able to escape to Portugal. At the time, one of the best-known illegal escape routes operated out of Marseilles, where the American Vice-Consul Hiram Bingham and the journalist Varian Fry helped refugees by raising money and bribing officials. Bingham and Fry secured exit papers and American visas for thousands, which included Arendt and her family. They first traveled through Spain to Portugal and from there sailed to New York City. Arendt taught at several institutions but predominantly at the New School for Social Research in New York City as an untenured professor. Alvin Johnson, Director of the New School, was one of the first higher education administrators to realize the danger Jewish and antifascist scholars were facing in Nazi Germany and Europe. At the New School, he founded a "University in Exile," and within a period of ten years, recruited 178 academics from Europe. Among them were art and film theorist Rudolf Arnheim, Austrian composer Hans Eisler, and theater director Erwin Piscator.

Although there are certain similarities between the respective knowledge transfer to the US and Turkey by the émigré scholars, the difference between living and working in a Western society and landing in a radically different culture and language limits the analogy. The German academics in Turkey were officially invited by the state and, therefore, unlike their colleagues or compatriots, did not have to navigate circuitous and often perilous routes to safety. They did not need to seek employment once they arrived at their destination, since they came with contract in hand. They were assigned to institutions where their expertise was needed and had a free hand in developing curricula. The academic infrastructure and niveau of knowledge in the respective host lands differed greatly. While the American universities already had qualified faculty in all disciplines and an established structure of higher education, in the Turkish case, the need was for a complex restructuring of the existing educational system that amounted to founding a modern university. On July 31, 1933, the old *Darülfünûn* (House of Sciences) closed down and reopened on August 1, 1933 as *İstanbul Üniversitesi* (University of Istanbul). Thus, in addition to imparting knowledge, the émigré professors were tasked with organizational assignments, involving the establishment of institutes and faculties and raising the existing disciplines—law, medicine, natural sciences, dentistry, social sciences, and literature—to a high scientific niveau. They trained generations of professionals and academics. Such a multitasking endeavor could not be realized without translation. This was not the case in the US, where the émigré professors, except for those teaching modern and ancient languages such as French, German, and Latin, had to teach in English (which only a few of them had mastered). At the Turkish universities, each professor had an assistant or a student, who performed a consecutive translation from German and sometimes from French into Turkish. The students then took notes in Turkish; however, a good number of them wrote these in the Ottoman Arabic script and not the Roman alphabet, which had been introduced in 1928 as part of Atatürk's modernization reforms, only five years before the arrival of the German professors. Thus, students of the émigré professors had already been subjected to a translational imperative (Seyhan, 2005).

Safeguarding knowledge: The imperative to translate

In spite of its challenges, translation was a medium that students embraced to ensure the afterlife of their professors' work. In fact, it is through translation into Turkish of high-quality academic lectures that we can now appreciate the intellectual rigor the German professors brought to the classroom. Ernst von Aster, a polymath philosopher, natural scientist, and psychologist who first left for Swedish exile in 1933 and then came to Turkey in 1936, taught History of Philosophy in the Faculty of Literature; Philosophy of Law in the Faculty of Law; and Methodology in the Faculty of Economics at the University of Istanbul. His multiple areas of expertise were further enhanced during his Turkish exile and expanded to include the philosophical legacy of the Turks in the History of Philosophy course (von Aster, 1937). Macit Gökberk, who was von Aster's assistant and translator, is recognized as one of the most prominent Turkish philosophers. Gökberk translated much of von Aster's lectures on the history of philosophy, *Vorlesungen über Geschichte der Philosophie,* as *Felsefe Tarihi Dersleri* (Lectures on the History of Philosophy, von Aster, 1943). A relatively recent article on the Istanbul University lectures of Ernst von Aster once again underlines how translation performs a decisive role in retrieving knowledge that may otherwise be lost to time. The article, "Ernst von Aster'in Çağdaş Felsefe Ders Notları" ("Ernst von Aster's Contemporary Philosophy Class Notes"), illustrates how von Aster's oversubscribed lectures, delivered during the academic year 1943–1944, were copiously taken in long hand and in Ottoman-Turkish script by Cahit Tanyol, a student in the philosophy department at the time. Years later, the lecture notes were discovered by a new generation of philosophy majors. Over the years, one or more individuals attempted to transliterate these into the Roman alphabet. The article, which includes the full text of the transliterated lectures and a short intellectual biography of von Aster, is a tribute to his interdisciplinary range and pedagogical gift (Özkul & Şahin, 2019). The stunning clarity of lectures that cover the history of modern philosophy from Descartes to the relation between Newtonian physics and Kantian philosophy all the way to Hans Driesch's Vitalism and Henri Bergson's Spiritualism makes them an indispensable reading for today's philosophy students.

Translation and the afterlife of scholarship in exile

The early years of the Turkish Republic were marked by the experience of a momentous transition and transformation, literally from one civilization to another. During this time, translation in both a literal and figurative sense became a key pillar of the architecture of modernization. The major reform acts of the era, such as the alphabet reform and the language reform, which involved replacing Arabic and Persian words with pre-Ottoman Turkish words and neologisms, and the university reform represented interlinked modalities of translational practice. Translation became not only a modus operandi of knowledge transfer and reform, but it also secured the *Nachleben* (afterlife; à la Benjamin) of scholarship undertaken in exile. In the US and, to a greater extent, in Turkey, academics who participated actively in translational activity by becoming translated agents themselves, that is, by lecturing and publishing in the language of the host country, left an enduring intellectual legacy. Hannah Arendt, Herbert Marcuse, and Erich Fromm, who settled in the US and published in English, became household names in academic and literary circles. Arendt's *The Origins of Totalitarianism* (1951), Fromm's *Escape from Freedom* (1941) and *The Art of Loving* (1956), Marcuse's *Eros and Civilization* (1964), *One-Dimensional Man* (1964), and *An Essay on Liberation* (1971), among their many other books, were academic and trade bestsellers, which are still in print, and their ideas remain relevant for the ongoing trials of modernity. On a personal note, Marcuse's *An Essay on Liberation* is a staple of my Readings in German Intellectual History course syllabus and remains a timely and relevant text that has continued to inspire my nineteen- and twenty-year-old college students.

The influence of the exiles in their respective host lands coincided with times of major historical transition in both Turkey and the US. Whereas this period corresponded roughly to the late 1930s and early 1940s in Turkey, the impact of the permanent exiles in the US was most pronounced during the Cold War and Vietnam War years, as Arendt, Marcuse, Fromm, and Leo Löwenthal participated in the then current history of their adopted land and published works that analyzed the historical roots of discontent, authoritarianism, populism, anti-Semitism, and the twilight of reason in the Western world. What the lives and careers of émigré professors in the US have shown is that those who adapted to linguistic and cultural change transformed the American academic and cultural landscape and continue to remain relevant in the sociocultural life of their adopted country.

The work of German professors in Turkish exile, who like their fellow émigrés in the US, mastered the language of their adopted land and became deeply involved in its cultural milieu, has left an enduring impact on Turkish educational, social, and even political institutions. Georg Rohde, Ernst Reuter, Ernst E. Hirsh, and Fritz Neumark demonstrated a herculean feat in mastering a language radically different from their native one. They continued to publish both in German and Turkish and occasionally in French. Alexander Rüstow, known primarily as the father of neoliberalism—he coined the term—stayed in Istanbul for twelve years but did not learn Turkish. Nevertheless, he established close bonds with his students, colleagues, and assistants, who translated his lecture notes into Turkish, which were then published in book form (Rüstow, 1939, 1944). He was able to participate in many international conferences and maintain a level of academic productivity that would not have been possible had he stayed in Germany. He acknowledged his debt to Atatürk's Turkey in the foreword to *Freedom and Domination* (Rüstow, 1980, xxiii). His three-volume *Ortsbestimmung der Gegenwart*, written in Istanbul, is arguably on par with Marcuse's or Fromm's oeuvres. Rüstow's son, Dankwart Rüstow, a political scientist who spent his childhood in Istanbul and was fluent in Turkish, became a renowned scholar of Middle Eastern and Turkish studies.[5] He had *Ortsbestimmung* translated in an abbreviated English version. The shorter version condensed the original 1,795 pages with hundreds of pages of footnotes into 752 pages. The abridged English translation was then retranslated back into German as *Freiheit und Herrschaft: Eine Kritik der Zivilization* (Freedom and Domination. A Critique of Civilization, Rüstow, 2006) and led to an ever-growing interest in Rüstow scholarship, as attested by a number of critical commentaries on his very timely ideas and the popularity of condensed anthologies of his works.[6]

The retranslation of a book originally published in German from its abbreviated English translation—with thousands of footnotes cut, edited, and streamlined—back into German once again brings up questions of minor versus major languages, translatability, accessibility, and the politics of translation. It is only several years after its English translation was issued by Princeton University Press in 1953 that *Mimesis* came to be regarded by critics as the towering achievement of Western literary criticism. After the 50th anniversary edition of the English translation was published in 2003, literary critics, including Kader Konuk (2010)[7] and Emily Apter (2003),[8] analyzed the book as the foundational text of humanist legacy and comparative literature, respectively. While *Ortsbestimmung* and *Mimesis*, both written during their authors' Istanbul exile, have enjoyed a healthy afterlife in translation, books of arguably similar importance in their respective fields by exiled professors in Istanbul and Ankara remain undervalued. I contend that multidirectional translations in knowledge transfer are of invaluable importance for research

---

5   Dankwart A. Rüstow (1924–1996) was a German American scholar, who had an illustrious academic career in the US. He was the author of several books, notably, *Turkey, America's Forgotten Ally* (New York: Council on Foreign Relations Press, 1989) and *Politics and Westernization in the Near East* (Princeton, NJ: Princeton University Press, 1956), and co-editor of a volume on comparative politics, *Political Modernization in Japan and Turkey* (Princeton, NJ: Princeton University Press, 1964). His last name is anglicized with the omission of the umlaut.

6   See, for example, *Herrschaft oder Freiheit. Ein Alexander Rüstow Brevier* (Domination or Freedom. An Alexander Rüstow Breviary, 2007).

7   Konuk's extensively researched and illustrated work has been instrumental in drawing critical attention to the work of German academic exiles in Istanbul.

8   See, for example, Apter's (2003) "Global *translatio*: The 'Invention' of Comparative Literature, Istanbul, 1933," which appeared around the same time as the anniversary edition of *Mimesis*, and Apter's *The Translation Zone: A New Comparative Literature* (2006), where she also discusses Edward Said's interpretation of Auerbach's *Mimesis*.

in exile. However, today, only translation into English, the modern lingua franca, insures the afterlife of a work.

Competing mandates in translation

Although translation has become a regulative and cosmopolitan modality of our time and of shifting borders and populations on the move, it has also caused anxiety about the economy of equitable exchange between dominant (high status) and "minor" languages. Milan Kundera, the multilingual Czech writer who has been living in Parisian exile since 1975 and publishing his later works in French, is keenly aware of the risk that lack of accessibility or translation poses to languages and cultures of "small" nations. The smallness, in his words, is not one of scale but of destiny, the destiny of nations that "have all, at some point or another in their history, passed through the antechamber of death; always faced with the arrogance of the large nations" (Kundera, 1993, p. 192). Kundera's specific examples of "the small European nations" are Central European nations, such as Czechoslovakia, Poland, or Hungary, but the concept applies to all nations "secluded behind their inaccessible languages," (p. 193) and whose cultures are excluded from international recognition. I believe that this exclusion is not so much a result of inaccessibility—Why should Czech be less accessible than Russian in a linguistic sense?—as it is that of a nation's geopolitical standing and, to some extent, the unavailability of a nation's language(s) in translation. While Kundera does not mention lack of translation as a reason for inaccessibility, he has been obsessive about his own work, especially his books in Czech, being properly translated.

The Turkish experiment in translation defies the conventional wisdom that the translation of a "minor" language (Turkish) into a high status one (German) ensures the former's survival. The modernization of the educational institutions via the mediation of the émigré scholars shows how translation from a major language such as German into Turkish preserved a banished intellectual culture and insured its afterlife and survival. The translation work of Turkish and German scholars in Istanbul and Ankara confirms Walter Benjamin's observation that in translation, both the source and the target language encounter one another on an equalizing plane, where each is reciprocally enriched and expanded (Benjamin, 1977, pp. 50–62).

Ankara: The new capital as locus of sociocultural sea change

Unlike their fellow exiles in the US, the German émigrés in Turkey were seen as architects of a new national culture, which was to be freed from its Islamic past. In fact, modern architecture itself became an emblem of the new nation's Westernization project. German architects in Turkey were commissioned to create a culture of architecture that defined Turkey's debut into the Western world. One of the reasons Ankara was chosen as the site of the new capital city was that unlike Istanbul, where the past is literally memorialized in Byzantine and Ottoman monuments, in sultans' palaces, and in magnificent mosques, it did not carry the burden of the past. It was a small provincial town, resting on a steppe and could be built from the bottom up. The German architects and city planners had a free hand in designing a modern capital city, where government buildings and educational institutions would reflect the ethos of a modern, secular new nation. Therefore, it is perhaps no coincidence that Ernst Reuter and renowned architects Ernst Egli, Clemens Holzmeister, and Bruno Taut, among many others, were called to Ankara.

There is neither time nor space to do justice to the larger narrative of a singular cultural transfer in the humanistic work of Ernst Reuter or Bruno Taut, which not only stamped Turkish higher education with the seal of an enlightened modernity but also intervened in political policy. Balancing the critical archive with a renewed evaluation of their intellectual and political contribution will bring the work carried out by many intellectual exiles in this century and the last into a more sustained and nuanced dialogue across communities of research. That is the subject of my larger work. I will only mention the significance of the work of Reuter and Taut briefly, whose respective works in urban planning and architectural design became their most enduring legacy.

The German signature in city planning and architecture

Ernst Reuter's life was a most exemplary one, not only in terms of his intellectual prowess and the educational miracle he performed at the University of Ankara, but also his unflagging commitment to the restoration of a new Germany after the war (Möckelmann, 2013).[9] Removed from his office of the mayor of Magdeburg and sent twice to the Lichtenberg concentration camp by the Nazis, he ultimately arrived in Turkey and first took positions in the Ministry of Transportation and then the Ministry of Economics in Ankara. Fluent in Russian, due to his imprisonment as a prisoner of war during the First World War, and other languages, he learned Turkish in record time and began lecturing and publishing in Turkish. He taught an interdisciplinary arrangement of subjects, including city planning, municipal affairs, and municipal finance. As a professor at the *Siyasal Bilgiler Yüksek Okulu* (School of Political Science) at the University of Ankara, he established the discipline of Urban Planning, where today the Ernst Reuter *İskân ve Şehircilik Araştırma ve Uygulama Merkezi* (Ernst Reuter Center for Research in Urban Settlements) carries on his work. His *Komün Bilgisi. Şehirciliğe Giriş* (Communal Study. Introduction to Urbanism), published in 1940, still counts as a foundational text of urban studies in Turkey. Although Reuter returned to Germany as soon as the war was over, his intellectual legacy still remains very much alive in the city, where more than eighty of his works, published during the years of his Turkish exile (1938–1946) are catalogued at the Ernst Reuter Center.

Bruno Taut was already a renowned architect before he arrived in Turkey, his last station of exile. Taut's magnificent *Dil ve Tarih-Coğrafya Fakültesi* (Faculty of Language and History-Geography) building at the University of Ankara and his design of the Ankara Opera, and Ernst Egli's İsmet Paşa Women's Institute, School of Aviation, and other educational institutions symbolized the dream of the modern nation. Taut, Egli and Clemens Holzmeister put their signature on buildings, designed to reflect a modern national identity in the new capital city. Although Taut died prematurely in 1938 after a two-year stay in Turkey, he was the principal architect of fifty schools and institutes, in addition to the Faculty of Language and History-Geography building. Today, this faculty is still one of the last strongholds of secular higher education. At the entrance, there is a memorial stone for Taut. The entrance was paved by using stone and brick together, a nod to the famous Ottoman architect Sinan, whose statue actually adorns the green field in front of the building. Taut's admiration for Sinan was so great that he claimed Sinan's Süleymaniye Mosque, from the perspective of harmony and beauty, was superior to the Hagia Sophia (Taut, 2007, p. 249). He was

---

9   For an extremely well-researched and arguably the best critical biography of Reuter, see Möckelmann (2013).

known to have a penchant for the architecture of the East. In August 1916, when he first sighted Istanbul, he exclaimed enthusiastically, "Der Orient ist die wahre Mutter Europas, und unsere schlummernde Sehnsucht geht immer dorthin" (The Orient is the true mother of Europe, and our slumbering longing always moves thereto) (Taut, 2007, p. 73). His enthusiasm reminds us of the early German Romantic Novalis's (Friedrich von Hardenberg) idealization of the Orient. However, his romanticized image was not a mere image. He saw in the architecture of Turkish mosques the realization or concretization of a philosophical sense of life.

A paper by Christoph Ehmann, the former *Staatssekretär* (Deputy Minister) of the German Federal Ministry of Education and Research, who has provided me with valuable information, interprets Taut's brilliant architectural designs less as an expression of his craft than a confirmation of his *unparteiisch* (impartial) yet politically left-leaning sympathies. His career at various stations of exile strongly confirms this view. He was first a *Revoluzzer*, which in the words of Erich Mühsam characterized those with socialist sympathies before the First World War, and then a reformer. As a marked leftist, he was "forgotten" for years and was only restored to architectural history as the West German public slowly freed itself from the reactionary tendencies that persisted for two decades after WWII (Ehmann, 2013, p. 2). Surprisingly, this humanist creator originally chose to go to Russia to realize his utopian vision, only to return to Germany in 1933 after a year without having undertaken a single project. Landing on the pogrom lists shortly after his return to Germany, he fled to Switzerland and then to Japan. However, the political circumstances in Japan, thirst for war, and an imperialistic fever were in total opposition to his beliefs. Despite the fact he was treated as a celebrity, he could not build but ended up having to design ashtrays, lamps, and small pieces of furniture (Ehmann, 2013, p. 24). He soldiered on because he loved Japan. Finally leaving Japan, he came to Turkey in 1936, where his dreams of building were fulfilled. He was finally able to design schools, which totally departed from the foreboding Prussian ethos and were bathed in light that he considered conducive to learning. He built feverishly, putting his signature on numerous cultural edifices. And in a sad parting gesture, even designed Atatürk's catafalque, shortly before his own death (Ehmann, 2013, p. 29).[10]

---

10   I thank Prof. Dr. Christoph Ehmann for giving me his unpublished lecture, held on December 24, 2013 on the occasion of Taut's 75th Death Anniversary, "Vertreibung aus Deutschland-Berlin 1933 und Exil in Japan und der Türkei bis zu seinem Tode am

From hamlet to *Hamlet*

I would like to note that the professors who went to Ankara did not enjoy the city, as it lacked the more urban and cosmopolitan atmosphere of Istanbul, not to mention the mild climate and natural beauty of the city astride two continents. While Auerbach and many other professors appointed to faculties at the University of Istanbul lived in the upscale neighborhood of Bebek on the Bosporus shore, those appointed to positions at the University of Ankara and government agencies found themselves literally in a village, where housing was scarce, the climate arid with cold winters and hot summers, and retail stores few and far between. On the other hand, the émigré scholars and artists in Ankara did not have to defer to "old school" academics, a history that had to be preserved, or a blueprint for any institute or foundation. Thus, a unique faculty like Language and History-Geography, an Urban Studies Center, and centers for innovative research were established without the benefit (or the detriment) of a precedent. Renowned names in the arts such as Carl Ebert, Paul Hindemith, and the legendary music teacher Eduard Zuckmeyer, who integrated the principles of German *Jugendmusikbewegung* into Turkish music pedagogy (Widmann, 1973, pp. 141–142)[11] transformed an Anatolian hamlet into a capital of fine arts. Zuckmeyer stayed in Ankara until the end of his life, even though his wife Gisela Jockisch and her daughter, whom he adopted, remigrated to Germany in 1950. His life, music, and permanent Turkish exile would be the subject of another book. Ebert, a prominent actor and stage director who had trained under Max Reinhardt, was a strong opponent of the National Socialists. Upon Hitler's seizure of power in 1933, he left Germany, and after a successful career at many stations of exile, he was invited in the fall of 1939 to Ankara by Atatürk. During his Ankara sojourn from 1939 to 1947, Ebert established the theater and opera schools at the Ankara Conservatory and trained some of the most well-known actor-directors of the Turkish stage, among them, Cüneyt Gökçer, a virtuoso interpreter of Shakespeare.

24.12.1938" ("Expulsion from Germany-Berlin 1933 and exile in Japan and Turkey until his death on December 24, 1938").

11   Zuckmeyer is the subject of a 2015 documentary, *Eduard Zuckmeyer – Ein Musiker in der Türkei* (Eduard Zuckmeyer-A musician in Turkey) by Barbara Trottnow. The documentary can be viewed on YouTube at https://www.youtube.com/watch?v=r-LV 7KWb664. *Jugendmusikbewegung* (youth music movement) was a pedagogical music trend in the early twentieth century. It was influenced by the youth movement and aimed to preserve traditional folk songs.

Georg Rohde: Architect of an East-West humanism

George Rohde is one of the few prominent émigré scholars whose involvement in the promotion of classical studies in Turkey, contribution to a universal translation project, and training of accomplished classicists and translators attest to his enduring legacy. In 1931, he became a lecturer in Latin at the University of Marburg, replacing Paul Friedländer, who before him was ousted by the Nazi regime and went on to become a distinguished faculty member at the University of California in Los Angeles. In 1935, Rohde was invited to assume a professorship of Classical Philology in the Faculty of Language and History-Geography in Ankara. He founded the Institute of Classical Philology and single-handedly built its library. The introduction of Latin lessons to the Turkish secondary school curriculum was a first, as well as the appointment of a foreign national to the post of superintendent of Latin instruction. Unlike the US, where Latin was regularly taught in high school, especially at Catholic schools, and where most universities had established departments of classics, in Turkey, where the use of the Roman alphabet was a relative novelty and students were still being weaned from the use of Arabic and Persian loan words, introducing Latin into the curriculum would have been a challenge for any educator. But Rohde was not just any educator.

Rohde's daughter Silvia Giese shared details of Rohde's family's life in Ankara with me.

Her father wasted no time in getting the family acclimatized to life in Ankara. Silvia and her brother Firman were sent to Turkish schools, where they became star students. With his assistant Samim Sinanoğlu, who became a leading Turkish classicist, Rohde co-wrote *Lingua Latina: Lâtince Ders Kitabı* (Lingua Latina: Latin Textbook) Part I (1948) and Part II (1950). The textbooks included selections from Caesar and Ovid and were reissued many times. Plato's *The Republic* was translated into Turkish by Rohde and his assistants at the Faculty of Language and History-Geography. Both Samim Sinanoğlu and his brother Sait, a scholar of Ancient Greek, were Rohde's assistants. They are among the most renowned scholars to emerge from Turkish academia. Another of Rohde's assistants was Ekrem Akurgal, alumnus of the University of Berlin and one of the most prominent Turkish archaeologists, who has been honored with the Goethe Medal and the *French Légion d'honneur Officier*, among many other national and international honors.

Azra Erhat, another assistant and co-translator of Rohde's, is now remembered as a writer and scholar who popularized classical studies and unearthed the histories of ancient Anatolian sites in her best-selling books. Erhat was a student of Leo Spitzer at the University of Istanbul, who recommended her to Rohde to translate his lectures from German and French as well as his works in Latin and Greek to Turkish. Rohde inspired his Turkish students to blend Western classics with the legacies of Ottoman-Turkish culture to create the basis for Turkish humanism. Erhat embodied this distinctive humanism by synthesizing Western and Anatolian myths and legends not only in her scholarly translations, but also in her popular books. Her linguistically and formally accomplished translations, which also reflected an Anatolian identity she cherished, remain a gift to Turkish readers, who otherwise would have had little access to the founding works of Western humanism. The objective of Suat Sinanoğlu's *L'humanisme à venir* (The coming humanism,1960) and his three-volume *Türk Hümanizmi* (Turkish Humanism, 1980), which is a philosophically grounded interpretation of Atatürk's modernization reforms, was to introduce "classical thought" as the foundation of Western civilization, not only to the Turkish nation, but also to cultures beyond the Christian world through the mediation of a new Turkish humanism. In *L'humanisme à venir*, Sinanoğlu maintains that "Atatürkism," in its physical as well as spiritual and intellectual aspects, was synonymous with a genuine Westernization that would raise society from the imitation of the modern to the application and implementation of the modern (Sinanoğlu, 1960).

Some of Rohde's major contributions to the establishment of Turkish humanism were his organization of conferences outside the university at various community centers, thus joining town and gown, and his participation in the translation projects of the series "Translations from World Literature" under the auspices of the Turkish Ministry of Culture. In this series, Rohde translated Plato's *The Republic*, Books I and II, and Azra Erhat translated Book III (Widmann, 1973, p. 286). Unlike some of his émigré colleagues at the University of Istanbul, who preserved the aura of the ivory tower professor, biding their time until an invitation came from greener Western pastures, Rohde remained committed to a mandate of "translating" genuine humanism into Turkish education, long after his original assignment was completed. His students and assistants, now also all gone, valiantly carried his legacy forward, educating successive generations of classicists. What still remains of this foundational "classical thought" of humanism under the current regime that is advocating a return to political, not humanist Islam, as Suat Sinanoğlu had envisioned it, is indebted to the work of

Rohde and his mentees. While Auerbach and Spitzer at the University of Istanbul were beloved mentors to a generation of Turkish philologists, their influence beyond the walls of academia remained limited due to a language barrier. Rohde, Reuter, and Zuckmeyer understood that communication is best achieved if language speakers meet halfway. They studied the host country's language and worked with translators, often as co-translators – a process that advanced their own knowledge of the target language – or translated their own works. In the now defunct Ankara newspaper *Ulus* in his daughter's archive, Rohde is shown with the Turkish President İsmet İnönü, supervising an exam. In fact, the Ankara cadre of exiled scholars was often in the news, as they were fluent in Turkish and were major cultural players in the capital city. I focus on the Ankara group to foreground the important role they played, like their American émigré counterparts, in real life or even the Realpolitik of the host nation. Auerbach was certainly a great mentor, even uncharacteristically close for a German professor to his students (Urgan, 1998, pp. 175–176).[12] However, *Mimesis* has to this day not been translated into Turkish. And that is a great loss for Turkish students and readers. Its fame rests on the original language and its translation into English. Auerbach and Spitzer taught French literature and the great works of Western literature, but unlike Rohde or Zuckmayer, they did not draw on the linguistic and cultural capital of the host country. It is arguably in Rohde, his students, and assistants' work at the University of Ankara that a humanistic legacy, which a young secular republic aspired to embrace, was most directly introduced into Turkish educational curricula. Thus, I maintain that Rohde helped create a Turkish humanism that reached beyond the book and Turkish borders and still endures under the antihumanist regime of today that keeps hundreds of writers and journalists in prison on fabricated charges of terrorism and treason.

---

12   See, Mina Urgan, *Bir Dinozorun Anıları* (Memories of a dinosaur, 1998). Urgan studied French philology with both Spitzer and Auerbach. She was a professor of English literature at the University of Istanbul and a human rights activist. Her memoir reveals Auerbach's very close bond with his students, with whom he went on skiing trips to Uludağ, near the city of Bursa, even though due to a permanent foot injury he suffered as a soldier in the First World War, he could not ski.

Excursus: Displaced scholars at traditionally Black colleges

As I mentioned in my introductory section, the story of the German academic émigrés, who taught at the traditionally Black colleges in the American South, needs to be remembered in the larger history of academic exiles. In my view, their life and work in American exile bears little, if any, resemblance to the privileged positions of their fellow German writers and scholars at prestigious institutions. In terms of their contribution to the social and intellectual life of their communities, they stand much closer to their compatriots in Turkish exile, who contributed much to the sociopolitical fabric of the host land. Their spirit and ingenuity in a culturally alien and even, to some extent, hostile environment may provide both inspiration and consolation to today's displaced academics. Thus, I hope to retrieve from historical obscurity the memory of these lesser-known names, who settled in the American South and valiantly participated in the Civil Rights Movement.

This group of refugee scholars included such noteworthy figures as philosopher and literary critic Beate Berwin, philosopher and legal scholar Ernst Borinski, historian Georg Iggers, philosopher and classicist Ernst Moritz Manasse, and Marxist philosopher and economist Fritz Pappenheim. Their stories bear witness to the resilience and resourcefulness of scholars committed to preserving and sharing a humanist legacy at all costs. Thanks to a little-known book, *From Swastika to Jim Crow* (1993) by Gabrielle Simon Edgcomb, a refugee from Nazi Germany herself, the names and remarkable life stories of these émigré scholars have been preserved for the archive of modern intellectual exile. When Jewish academics were forced to flee Nazi Germany, American institutions embraced only well-known names with connections in high places, like Albert Einstein, Hannah Arendt, and Theodor Adorno. Lesser-known figures, though lucky to have escaped to the US, struggled to gain a foothold, not only because of the Great Depression, which had depleted jobs, but also because of the prevalent anti-Semitism and anti-German sentiments. Although the traditionally Black colleges, Howard University in Washington, D.C., Lincoln University in Chester County, Pennsylvania, Fisk University in Nashville, Tennessee, and Hampton Institute in Hampton, Virginia, among others, could not offer the émigrés a prestigious address, they welcomed them as friends in fate. The common experience of persecution bonded the deposed German-Jewish scholars with the Black Americans and created a haven of interracial, intercultural, and intellectual comradeship. Many émigré scholars, such as Georg Iggers and sociologist

Ernst Borinski, who taught at Tougaloo College, Mississippi, became active in undermining Jim Crow laws. Before arriving in the US, Borinski had also earned a doctor of philosophy degree at the University of Berlin in 1928 and an international law degree from the Academy of International Law in The Hague, Netherlands in 1930. Georg Iggers, who became one of the most distinguished scholars of European history, fled Germany with his family only a few weeks before *Kristallnacht* (Night of Broken Glass). He taught for several years, first at the Philander Smith College in Little Rock, Arkansas, and then at Dillard University in New Orleans. In 1951, he and his wife joined the National Association for the Advancement of Colored People (NAACP). He played an active role in efforts to desegregate public schools in Arkansas. After a committed career as educator and activist, Iggers was appointed as a faculty member at the State University of New York at Buffalo in 1965 and named Distinguished Professor there in 1977. In the words of the late John Hope Franklin, James B. Duke Professor of History at Duke University, the German-Jewish professors were "a remarkable group of men and women," who reminded "us that in the human family there are those who can transcend the Nazi swastika as well as Jim Crow, and set an example of human relations that their students and colleagues would do well to emulate" (Franklin, 1993, p. xi). Like their fellow émigrés in Turkey, these scholars found themselves in an unfamiliar world yet put their vast knowledge and interdisciplinary teaching skills at the service of the host institutions.

Because there was a shortage of qualified teaching staff both at the Turkish universities and in the traditional Black colleges in the American South, the émigré professors, who found teaching positions at these institutions, taught not only subjects in their fields but in related as well as different fields. In addition to chairing the Sociology Department at Tougalou College, Borinski taught German and Russian. He was also a recognized expert in constitutional law. Beate Berwin had earned a doctorate in philosophy from the University of Heidelberg with a dissertation entitled *Das Unendlichkeitsproblem in Schellings Ästhetik* ("The Problem of the Infinite in Schelling's Aesthetics") as well as a *Dottore phil.* in literary studies from the University of Bologna in Italy (Sohst, 2003). At Bennett College in Greensboro, North Carolina, where she spent her whole career, she also taught in addition to philosophy, German and geography. Her critical monographs on Moses Mendelssohn, Friedrich Hölderlin, and Heinrich von Kleist were published in a collected edition in 2003, 33 years after her death (Berwin, 2003). The theoretical rigor as well as the sociohistorical contexts of these monographs insure their enduring critical relevance. Even though her work is arguably on par with or superior to that of her fellow academic émigrés, such as Liselotte and Herbert Dieckmann, Frieda Wunderlich, and Willie Rey, who found employment at major research universities, she remains an unknown name in intellectual history. Her Hölderlin essay is required reading in my German Intellectual History class. I maintain that the absence of a *Nachleben* for her work – as is the case with the works of other accomplished émigré scholars – is a consequence of its unavailability in English translation and not of her employment at Bennett rather than at Yale or Berkeley.

Conclusion

Although the subjects of this essay were for the most part fortunate to have found refuge in host lands, their lives were far from perfect, albeit in different ways. Andrea Deciu Ritivoi's *Intimate Strangers: Arendt, Marcuse, Solzhenitsyn, and Said in American Political Discourse* recounts the somewhat marginalized situation of the four "intimate strangers," who despite their respected scholarship and best-selling books in English, were always considered "foreigners," since among other factors, their accents always gave them away as such (Deciu Ritivoi, 2014). The German scholars in Turkish exile, on the other hand, were admired precisely because they were foreigners, since they represented the other, cherished Western way of life. Their mandate was to set the cultural course of a secular nationalist agenda, when their own lives were disrupted by a nationalism run amok. Despite the sad irony of their task, their collective work emerges as a meditation on the emancipatory possibility of an enlightened modernity against the backdrop of Theodor Adorno and Max Horkheimer's scathing depiction in the first chapter of *Dialektik der Aufklärung* (*Dialectic of Enlightenment*) of the wholly enlightened world "radiating triumphant disaster" ("*strahlt im Zeichen triumphalen Unheils*") (Horkheimer & Adorno, 1969, p. 9). While Auerbach remained skeptical of an incipient Turkish nationalism, academics such as Neumark, Rohde, Zuckmeyer, and especially Ernst Hirsch, who became fully conversant in Turkish culture and politics, understood the historical necessity of establishing some sense of national unity in a new nation state built on the ruins of a multiethnic empire that in its final years of decline had become ethnically, culturally, and religiously very divided. Like Jürgen Habermas after them, they regarded the Enlightenment as an unfinished, ongoing project.

Unlike Adorno, they saw in the ideals of the Enlightenment the possibility of rehabilitating the inheritors of the "sick man of Europe," as the Ottoman Empire came to be known during its decline.

In the final analysis, the exile of German academics in Turkey and the US shows how a certain order of cultural configurations acquires meaning in translation in the broadest sense and by its prospective arrangement. A series of conjectural events in the US, that is, the Civil Rights Movement, the assassination of John F. Kennedy, the Vietnam War, and student unrest and shootings, created the need not only for the lessons of the German émigré professors' experience of traumatic memory but also their ways of confronting and overcoming it in the power of language, as they translated agony into corrective remembrance. What began as an exodus of German exiles to Turkey led to a radical transformation of the university system, which in turn became a symbolic fortress of secular modernization. Under the constant threat of political Islam and despite being overrun by administrators and rectors appointed by the Islamist government, the university reform undertaken by the German émigré professors endures as the symbol of Turkey's once successful debut into Western modernity.

In our day, academics leave or are forced to flee their homelands for different reasons, including censorship, persecution, escape from oppressive regimes, and simply for personal and academic freedom. What we learn from the experience of exiled academics in Turkey and the US confirms what Rüstow wrote in the foreword to *Freedom and Domination*. He made the point that to make sense of the cataclysmic events of an age –referring to the trauma of the Third Reich – one discipline is insufficient (Rüstow, 1980, p. xxiv). While he did not use the as yet uncoined term interdisciplinarity, that is what he advocated and practiced. Exile demands retooling of resources. Teaching across disciplines and revising extant knowledge are modalities of translation. Thus, drawing on the critical and empirical issues that frame this volume, I posit two questions that may not be answered any time soon but urge serious contemplation: How can translation, broadly conceived as a transport of ideas, technology, images, and information, safeguard intellectual capital that is threatened, censored, silenced, and banished? and Can historical strategies to safeguard intellectual inheritance be repeatable in the face of current crises of awakening neofascism, surveillance states, financial inequality, and precarity? While we cannot answer these questions with any certainty, crises provide access to history and a working framework for investigation.

References

Apter, E. (2003). Global *translatio*: The 'Invention' of Comparative Literature, Istanbul, 1933. *Critical Inquiry, 29*(2), 253–281.

Apter, E. (2006). *The Translation Zone: A New Comparative Literature*. Princeton: Princeton University Press.

Auerbach, E. (1946). *Mimesis. Dargestellte Wirklichkeit in der abendländischen Literatur*. Bern: Francke Verlag.

Auerbach, E. (1953, 2003). *Mimesis: The representation of reality in Western literature* (W. R. Trask, Trans.). Princeton: Princeton University Press.

Benjamin, W. (1977). Die Aufgabe des Übersetzers. In W. Benjamin, *Illuminationen* (pp. 50-62). Frankfurt am Main: Suhrkamp.

Berwin, B. (2003). *Mendelssohn. Hölderlin. Kleist. Drei Monographien zur Einführung*, (W. Sohst, Vorwort). Berlin: Xenemos Verlag.

Deciu Ritivoi, A. (2014). *Intimate strangers: Arendt, Marcuse, Solzhenitsyn, and Said in American political discourse*. New York: Columbia University Press.

Ehmann, C. (2013). Bruno Taut zum 75. Todestag. Vertreibung aus Deutschland-Berlin 1933 und Exil in Japan und der Türkei bis zu seinem Tode am 24.12. 1938. (Unpublished manuscript).

Franklin, J.H. (1993). Foreword. In G. Simon Edgcomb, *From Swastika to Jim Crow: Refugee scholars at Black Colleges*. (pp. ix–xi). Malabar, FL: Krieger Publishing Company.

Hirsch, E.E. (1982). *Aus des Kaisers Zeiten durch die Weimarer Republik in das Land Atatürks*. München: J. Schweitzer Verlag.

Horkheimer, M., & Adorno, T.W. (1969). *Dialektik der Aufklärung. Philosophische Fragmente*. Frankfurt am Main: S. Fischer Verlag.

Konuk. K. (2010). *East West Mimesis: Auerbach in Istanbul*. Stanford: Stanford University Press.

Kundera, M. (1993). *Testaments betrayed; An essay in nine parts* (Linda Asher, Trans.). New York: Harper Perennial.

Möckelmann, R. (2013). *Wartesaal Ankara. Ernst Reuter-Exil und Rückkehr nach Berlin*. Berlin: Berliner Wissenschaftsverlag.

Neumark, F. (1980). *Zuflucht am Bosporus. Deutsche Gelehrte, Politiker und Künstler in der Emigration 1933-1953*. Frankfurt am Main: Knecht.

Özkul, O., & Şahin, Y. (2019). Ernst von Aster ve Çağdaş Felsefe Tarihi Notları. *Beytulhikme. An International Journal of Philosophy, 9*(I), 231–277.

Reuter, E. (1940). *Komün bilgisi. Şehirciliğe giriş [Kommunalwissenschaft. Einführung in das Städtewesen]*. Ankara: SBYO.

Rohde, G. (1948, 1950). *Lingua Latina. Lâtince Ders Kitabı* (2 Vols.) (S. Sinanoğlu, Trans.). Ankara: MEB.

Rüstow, A. (1939, 1944). *İktisadi Coğrafya* (Vol. 1, R. Ş. Savla, Trans.; Vol. 2, H. İlteber, Trans.). İstanbul: İstanbul Üniv. İktisat Fakültesi.

Rüstow, A. (1950, 1952, 1957). *Ortsbestimmung der Gegenwart. Eine universalgeschichtliche Kulturkritik* (3 Vols.). Erlenbach-Zürich & Stuttgart: Eugen Rentsch Verlag.

Rüstow, A. (1980). *Freedom and domination. A historical critique of civilization* (S. Attanasio, Trans.). Princeton: Princeton University Press.

Rüstow, A. (2006). *Freiheit und Herrschaft. Eine Kritik der Zivilization.* Münster: LIT.

Rüstow, A. (2007). *Herrschaft oder Freiheit. Ein Alexander Rüstow Brevier.* (Michael von Prellius, Ed.) Bern: Ott Verlag.

Seyhan, A. (2005). German academic exiles in Istanbul: Translation as the *Bildung* of the Other. In S. Bermann & M. Wood (Eds.), *Nation, language, and the ethics of translation* (pp. 274–288). Princeton: Princeton University Press.

Simon Edgcomb, G. (1993). *From Swastika to Jim Crow: Refugee scholars at Black Colleges.* Malabar: Krieger Publishing Company.

Sinanoğlu, S. (1960). *L'humanisme à venir.* Ankara: Ankara Üniversitesi, Dil, Tarih- Coğrafya Fakültesi.

Sinanoğlu, S. (1980). *Türk hümanizmi.* Ankara: Türk Tarih Kurumu.

Sohst, W. (2003). Vorwort. In B. Berwin, *Mendelssohn. Hölderlin. Kleist. Drei Monographien zur Einführung.* (pp. III–XIV). Berlin: Xenemos Verlag.

Taut, B. (2007). Reiseeindrücke aus Konstantinopel 1916. In M. Speidel (Ed.), *Ex Oriente Lux. Die Wirklichkeit einer Idee* (pp. 73–78). Berlin: Gebr. Mann Verlag.

Urgan, M. (1998). *Bir Dinozorun Anıları.* Istanbul: YKK.

von Aster, E. (1937). *Die Türken in der Geschichte der Philosophie.* Istanbul: Devlet Basımevi.

von Aster, E. (1943). *Felsefe tarihi dersleri* (M. Gökberk, Trans.). Istanbul: Ahmet İhsan Matbaası.

Widmann, H. (1973). *Exil und Bildungshilfe. Die deutschsprachige akademische Emigration in die Türkei nach 1933.* Bern: Herbert Lang & Frankfurt am Main: Peter Lang.

Redefining precarity through knowledge production
The experience of the Indonesian 1965 exiles

Rika Theo & Maggi W.H. Leung

Kuslan Budiman (1935–2018) was an Indonesian theatre artist and writer who studied theatrical stage design in China and spent more than half his life living in exile. In 2018, he passed away at a hospice in the Netherlands, the last country he resided in after decades of living precariously as a stateless man in China and Russia. Because he was an office-bearer of a left-leaning organization, namely *Lembaga Kebudayaan Rakyat* (Lekra; Institute of People's Culture),[1] Kuslan's citizenship was revoked by Indonesia's anti-communist New Order government in 1966, and like many of his compatriots, he could not return home. Kuslan lived alone for 27 years in the Dutch city of Woerden, but his last days at the hospice were filled with visitors, mostly his exiled friends as well as Indonesian activists, scholars, and students. These were the people with whom Kuslan engaged as he continued his activism while living in exile. "Eksil[2] should realistically engage with the movement," he once said (K. Budiman, personal communication, February 11, 2017). The movement he was referring to was engaged in a long-term struggle to reveal injustice and violence in 1965, which is still presented ambiguously in Indonesian official history. As a survivor of this injustice, Kuslan wrote about his experiences in various articles and poems published in Indonesia and the Netherlands. Although he did not work formally as an academic, he was actively involved in scholarly and public discussions both offline and online, occasionally hosted Indonesian students, activists, and journalists at his home, and he was a

---

1 A literary, art, and social organization associated with the Communist Party of Indonesia.
2 The Indonesian term for people living in exile, especially referring to those exiled as a result of the 1965 communist extermination.

member of the art community group in his Woerden neighborhood. As such, we consider him an "academic intellectual in exile."

Kuslan was one of the many Indonesian intellectuals in exile who, despite the hardship of living in precarity, stayed connected to their "home" – Indonesia – and the diaspora through sharing their thoughts and exilic experiences. He was one of the many hundreds of students, party cadres, and people with potential intellectual talents who were sent abroad to study by the Sukarno government in the 1960s as a part of a nationalistic state project and growing cooperation with socialist countries. There is no official record of the precise numbers of these exiles, but according to Lebang (2010), in the 1960s, there were 2,000 Indonesian students, the largest international student body, in the USSR. Others offered a rough estimate of 700–800 students overall (Hill, 2014). The figures given by five respondents in this research varied from 60 to 100 Indonesian students in China alone. In addition, small numbers of students were sent to several socialist countries in Eastern Europe.

Among the gradually increasing amount of research on Indonesian exiles in the last decade, several studies explored how the Indonesian 1965 exiles reclaimed their agency while living in exile. Chambert-Loir (2016) and Setiawan (2010) analyzed *sastra eksil* (the exile literature), namely, the Indonesian 1965 exiles' collection of extensive writings on exile identity and experiences, their particular knowledge, and their multiple stances on the home country. Further, Nadzir (2018) examined the practices of agency in exiles' interconnected offline and online activism. Gurning (2011) investigated the exiles from the gender perspective, while Hearman (2010) studied these exiles in Cuba, Hill (2014) focused on them in the Soviet Union, and Mudzakkir (2015) concerned himself with those in the Netherlands. Much of the research highlights the exiles' active participation in long-distance nationalism.

Our chapter builds on this body of work. We examine how these scholars in exile navigated their lives that were diverted from privilege to precarity; map out their impact on knowledge transfer, exchange, and production even without an academic appointment; and analyze their lifelong efforts to share and connect with generations of Indonesian students and scholars as well as with non-Indonesians and others outside academic circles who crossed their paths. These scholars in exile have passed on to the younger cohorts "lost" collective memory, namely, an alternative narrative of Indonesian politics, history, and society that has hardly ever been mentioned in the official history of Indonesia, either during the authoritarian regime or after it.

This chapter draws on broader research on academic mobility between Indonesia and China over the past decades. Here, we draw on our data from fieldwork conducted in Indonesia, China, the Netherlands, and Sweden from 2016–2017. The first author (Rika, Indonesian) collected biographical narratives from five academics in exile[3] (four males and one female in their seventies or eighties at the time of interview). All of them had studied in China in the 1960s. The interviews were conducted in different cities and usually lasted more than three hours. In Sweden, Rika stayed with one of the interviewees, followed him as he went about his daily activities, and attended a meeting with him and other exiles. We also consulted literature on and biographies of the exiles to supplement the interviews. In 2017, a semi-structured focus group interview was conducted with eight current Indonesian students in the Netherlands who had come into contact with the elder scholars in exile.

In the following, we present the life stories of our interviewees and reflect on their impact on knowledge transfer, exchange, and production. We show how these scholars in exile have lived with different forms of precarity and redefine them in creative and resilient ways. Our findings highlight the temporally and spatially unbound nature of knowledge mobilities and production. Finally, we conclude with the implications of our research in respect to reconceptualizing the core notions we are dealing with in this book, namely, academics, knowledge, and exile.

The ruptured lifepaths and precarity of Indonesian exiles

The majority of the Indonesian scholars in exile were taken by surprise when they ended up in exile during their studies abroad. As *mahid* (the abbreviation of *mahasiswa ikatan dinas*, which means "students with a government bond scholarship"), they were put on a prescribed path to gain new knowledge abroad and then return to Indonesia and use that knowledge to help develop their country. These privileged future elites were instructed to study seriously, which also meant not marrying local people (Sipayung, 2011), as marriage might have prevented their return. This promising path suddenly ruptured

---

3   Sarmadji, Kuslan Budiman, Tom Iljas, Melanie, and Arkan were interviewed separately. This chapter uses the real name only for Sarmadji, Kuslan Budiman, and Tom Iljas, who have openly published their identity and activities in the media.

in 1965 when the "September 30 Movement", a secretive group of left-wing military officers kidnapped and killed several right-wing army generals (Eickhoff et al., 2017). This failed coup was blamed on the *Partai Komunis Indonesia* (PKI; Communist Party of Indonesia). Resulting from this, anti-communist mass violence, backed by the army, conducted in various forms took place (Robinson, 2017). About half a million Indonesians were killed and another million were imprisoned without trial. This destroyed the social base of Sukarno's presidency (Eickhoff et al., 2017). At the height of the Cold War, the Indonesian army under General Suharto seized power and established the New Order regime. The army's power under Suharto extended to Indonesia's embassies: Indonesian students[4] in China and other socialist countries were summoned to their local embassy and asked to sign a letter denouncing the previous government and submitting to the new one (see Hill, 2020 for the latest detailed research on the dilemma faced by the Indonesian exiles in China). Refusing to sign the letter meant they would have their citizenship revoked and thus be made stateless.

More than half of the Indonesian students in China chose not to return home, but then had to decide whether to stay in China with the support of their hosts or seek ways to move to other countries (Hill, 2020, p. 346). Either way, they were transformed from somebodies into nobodies. The upward social mobility promised to them became instead a trajectory of downward mobility. These students were forced into displacement from their home country, from the student mobility and academic achievements they had aspired to, and from their friends, families, and previous lives. They commonly experienced a feeling of defeat, loss, bitterness, resentment, and frustration due to a sudden disconnection from the personal and political projects they had built (Cornejo, 2008).

For the *mahid*, who were often dubbed Sukarno's students (Dragojlovic, 2012), and PKI cadres, returning home was not an option due to their ideals and fears. Stories about the slaughter and imprisonment of leftist and PKI sympathizers at home increased their fears and uncertainty. Many students could not return home and entered a precarious phase in regard to their movement from one country to another. Some managed to escape with the

---

4   In addition to the students, others were stranded abroad following the 1965 putsch, for instance the 500 Indonesian cultural delegates who were in Beijing to attend the anniversary of the Chinese Communist Party (Hill, 2010).

help of friends, family, fellow exiles, and activists, but continued living in uncertainty.

Melanie, the daughter of a 1965 political prisoner who studied medicine in China, parted from her husband, Arkan, during the difficult situation in 1965. Arkan, an Indonesian student from West Sumatra, whom she had married while they were studying in China, went to Hong Kong to work. After losing touch with him, she decided to leave China and finally arrived in the Netherlands with the help of an old friend of her father. She recounted this to us:

First, I went to Germany using my Indonesian passport. At that time, Indonesians were able to live in Germany without a visa for three months. I was worried about my passport, but fortunately, I was safe. Then, a Jewish lawyer in Belgium, my father's friend, said he was willing to sponsor me. He was the one who helped me with my invalid passport to get a Benelux residence permit. (Melanie, personal communication, December 1, 2016)

Staying for a while in one or more countries before eventually settling down was the norm for most of the exiles. Some traveled to Vietnam, Laos, Cambodia, or Myanmar to try to find a way to get home, but could not find one. Others went to the socialist countries that suited their ideology or to European countries that accepted political refugees.

However, they were not simply free to decide to leave or stay somewhere. In addition to needing an opportunity and the resources to enable them to move to a new place, they also had to change the route they took due to political changes in the place they thought would be their safe haven.

Kuslan experienced a series of unexpected changes along with the changing of places and situations. In the early 1970s, he was among those who were finally given a ticket to leave Beijing by the Chinese government:

I intended to go to the West, to Germany, but I stopped off in Moscow. I had a place to stay in Germany, but I did not know what I would do. I did not have much money, only 50 dollars. I could not exchange my money in China as I had defied the Delegation.[5] Someone even said I was a traitor for wanting

---

5   Indonesians related to the PKI's network in Beijing and Moscow did not wait passively. In February 1966, the Delegation of the Central Committee of the PKI was established in Beijing. It claimed to be the authoritative representative of the party and called all Indonesian leftists who were stranded abroad to go to China. Those in Moscow also founded the Foreign Committee of PKI with the support of the Soviet Communist Party (Chambert-Loir, 2016). This division was rooted in the further separation of Communist

to leave. In Moscow, my friend assured me that I would be better off staying. I was able to continue my studies even though I was already 37 years old at the time.

Kuslan became quite a prominent painter in Moscow. However, he left everything behind again when the Soviet Union ceased to exist in 1991. He arrived in the Netherlands and lived on benefits from the Dutch government.[6] Like Kuslan, many of the exiles, most of whom were born in the 1930s or 1940s, were middle-aged or older when they finally settled down. Age became one of the obstacles to finding jobs, especially those that called for academic qualifications. As they moved around without concrete ideas about their future, they had to take any opportunities that came along and improvise to make the best of things. In many cases, their previous skills and knowledge were not appreciated due to the limitations associated with the status of political refugees or migrants in general. Most of them ended up doing blue-collar, low-wage jobs, although some were able, after much effort, to further their careers in other areas of work.

For instance, Tom, who had an agricultural engineering degree from China, arrived in Sweden with his wife and children after spending three years in exile in Romania. Tom went to a Swedish automotive company to apply for a job. At the HR department, he was not asked about his previous education and skills, and was given a job as a factory worker whose main task was to turn screws.

> It was mechanical work, which I did easily and quickly. As I often helped other divisions, I drew attention, and then they gave me further training and promoted me to supervise the production division. During the training, I happened to correct a mistake by the trainer. The trainer was amazed and asked me how I knew it. He said that I should not have been in that class, as it was only for those who did not finish high school. He reported me to my supervisor who then asked me about my previous education. I just knew that there were two kinds of HR for a job interview. I was brought to the HR for manual work because of my appearance. Even though at that time I wore

---

China and Soviet Russia, with the PKI in its final years inclining towards China. This was one of the polarizations that happened among the exiled scholars.

6   The exiles who were born prior to Indonesia's independence in 1945 can be considered by the Netherlands as citizens of the Dutch East Indies (the Netherlands colonial state).

a suit, I still had black hair. (T. Iljas, personal communication, September 18, 2016)

Although Tom might have lost his initial chance to get a job in line with his skills due to the different treatment given to migrants, his skills and knowledge eventually enabled him to get a better job. He then became a trainer for the other production workers and, after 10 years, was promoted to head of a logistics department.

Difficulty finding a suitable job is a common story among exiles. As mentioned by the interviewees, age was a crucial factor. For example, Sarmadji was 45 when he arrived in the Netherlands in 1976. He had studied child pedagogy in Beijing, and during the precarious days in China after 1965, he was active in the production of *Suara Rakjat Indonesia* (*The Voice of Indonesian People*), a student news bulletin about Indonesia and the leftist movement in the years following the events of 1965. As he initially did not read or speak Dutch, he was given a job as a glasscutter laborer, without being asked about his education or training. He learned Dutch by memorizing words while working and was helped by three co-workers, Dutch ex-soldiers who had served in Indonesia.

In addition to age, we found that another obstacle to finding suitable jobs was the exiles' academic degrees: they were issued in socialist countries and generally not recognized in Western Europe. Few exiles found work in their field of study. However, a small number eventually worked their way up to become prominent academics in their host country. But even for those who had adjusted well, the process of doing so was full of uncertainty and difficulty. They often had to hold down several precarious jobs, experience downward socioeconomic mobility, and reeducate themselves before they managed to get a decent job.

Precarity in an economic or material sense leads to precarity in other realms of life, for instance, in building personal relationships. For some, the insecurity of life, the uncertainty about the future, and the lack of a fixed place to settle down imposed constraints on their personal and family lives. Sarmadji reflected on his particular situation:

I am married to books [laughs]. Well, I am not a homosexual. I am a straight man who likes women. The situation was just impossible for me to have a family. And here the culture is different and for me, it is difficult, and more, with the kind of job and hobby I have. (Sarmadji, personal communication, April 20, 2016)

Nevertheless, Sarmadji tried to build a life that was optimistic, as reflected by his chosen pseudonym. It is common for Indonesian exiles to have other names, and Sarmadji chose the name Wardjo – an abbreviation of the Javanesse words *waras* (sane) and *bedjo* (lucky). Wardjo is the reflection, hope, and reminder of his life.

Many exiles were also deprived of other relationships, but particularly painful for many was being cut off from their home country. During the 32 years of the Suharto regime, the exiles had almost no contact with the Indonesian state or its embassies. The state-imposed communist label and their stateless condition were used to legitimize the othering treatment by the state. The fervent desire among many of them to return home was difficult or impossible to fulfill until the 1980s. Not only was it very complicated for them to get a visa, but there was also a safety concern with regard to them and their families because the suspicion and stigma of being communists (and therefore being atheists and traitors to the nation) were still very strong. We heard stories of several exiles who were refused entry to Indonesia as well as stories of people who did manage to return. Tom recalled that he was among the first of the exiles to visit Indonesia because he had a relative in the military who guaranteed his safety (T. Iljas, personal communication, September 18, 2016).

This situation improved in the post-Suharto period. In 1999, the Indonesian president, Abdurrahman Wahid, personally apologized for the 1965 tragedy and publicly declared his desire to revoke the parliamentary decree banning the PKI and Marxism/Leninism in Indonesia. The parliament rejected his proposal and since then no other president has attempted to get the decree revoked (Chambert-Loir, 2016). Wahid also invited the exiles – whom he called "the wanderers" – to return. The Indonesian Minister of Justice was sent to the Netherlands to discuss matters with the exiles, including the possibility of reinstating their citizenship. At first, it was appreciated as a good initiative and created much expectation among the exiles, but then there were some critical reactions as they felt they had been positioned as guilty illegal migrants (Mudzakkir, 2015). Many exiles viewed the offer to return as mere rhetoric because the government ignored the core of the problem by not officially admitting the unjust policy and not proposing to rehabilitate the exiles. These core issues were not handled either by any follow-up from the minister or further discussion about exiles by any subsequent Indonesian president.

The exiles continue to this day to wait for their endeavors to bear fruit. For many, however, the wait persisted till they passed away while in exile. Even though the exiles are now free to visit Indonesia, they continue to be cautious. Tom, for instance, was deported by the Indonesian government when he visited his hometown in Sumatra in 2016. He was about to visit the mass grave of 1965 victims, among whom was his father, when he was arrested by the local police. Tom's story reveals that the anti-communist sentiment is still alive and that therefore the exiles may have to wait much longer to be rehabilitated. To return or not to return is not the only question, as their precarity extends to the uncertainty regarding when the state might publicly and officially affirm their innocence and guarantee their security if they return.

Redefining precarity by telling and sharing "lost" memories

Edward Said (2002) described being an exile as one of the saddest fates, continuously living in a marginal and anomalous life, with the stigma of being an outsider. An exile's life is precarious, as it is subject to instability and endangerment (George, 2016). The sudden turn of our interviewees' lives into uncertainty, vulnerability, insecurity, and disconnectedness is in stark contrast to their previously well-planned and privileged lifepaths. Not only do exiled scholars carry psychological baggage – a sense of defeat or guilt for having left dead, jailed, or disappeared comrades behind – they have also seen their dreams destroyed, families torn apart, careers ruined, and personal space restricted as they were forcefully displaced from their homes (Wright & Oñate Zúñiga, 2007).

Political precarity breeds further precarity within the lives and the existence of the exiles. They live in a new home but are haunted by questions, disappointment, and curiosity about their old home. Living in half involvement and half detachment, they exist in a median state (Said, 2002). The scholars in exile are displaced from their country of origin, but for the rest of their lives, they are unable to detach their minds and also to some extent their activities from that country. This ongoing connection with the place of origin is the crux of exiles (Roniger, 2017). Yet, this burden is often their source of creativity in engaging with precarity during their exilic journey. Since they cannot follow a prescribed path, what they do as intellectuals has to be invented (Said, 2002).

Various opportunities to engage with precarity present themselves during the exilic journey. Variation in place and time also contributes to the different kinds of responses to their precarity, and thus we need to realize that there is no singular exilic condition and experience. In several cases, exiles could be muted, passive, and politically immobile (e.g., see Hsiau, 2010 on post-war Chinese intellectuals in Taiwan). In other cases, they could remain agents of their own destiny and reclaim revoked national identity and citizenship, given that living in exile provides them with the opportunity to change statuses, upgrade skills, discover strengths, and develop new relationships (Roniger, 2017). This is what Roniger called the expanding effects of exile, the opposites of the constraining effects of exile. For example, Chinese intellectuals with the status of political exile in Western countries could obtain access to foreign knowledge, expand contacts with foreign governments and human rights organizations, and generate international publicity (Ma, 1993).

The five scholars in exile discussed in this chapter were disconnected from their home country. Yet, they chose not to detach themselves from it. After being displaced, they had to start over again many times and build new connections in these unfamiliar places where they may also have felt disconnected. We observed in this research that precarity among the exiles arises not only from the uncertainties in livelihood, but also from their identity and endeavors as exiles. Our conversations made it clear that they also perpetually experience precarity in their actual or desired connections while being disconnected from their homeland. While not agreeing with or despising the state regime, they realize that they can do little to directly change it. They live in tension between the yearning to connect and return (in some ways), on the one hand, and hesitation and attempting to detach, on the other hand. The feeling of connection in disconnection has become the basis for their engagement with precarity.

Sarmadji has found meanings in precarity and used it for knowledge creation work. He has focused on his passion for books and the silenced Indonesian history. In his small, state-subsidized apartment, he has built a library that is regarded as a complete library on Indonesian 1965 history. He is surrounded by thousands of books and is often visited by Indonesian students and researchers. He admits that it is also all he can do to continue to struggle for Indonesia as an exile.

Sarmadji's library symbolically reflects his ongoing connection to Indonesia despite his other life in the Netherlands. Although the majority of his books are on 1965 and left-wing ideology, he also collects books and material on other significant events in post-1965 Indonesia. For instance, he offered the first author of this paper documents on the rape of Chinese-Indonesian women around 1998 during the mass riots in Jakarta. Furthermore, collecting the books and opening his library connect him to the other exiles and leftist activists, as well as to current Indonesian students and researchers.

Like many of the exiles, he joined the exile community and frequently participated in discussions related to Indonesia's present situation or history and the exiles themselves. In addition to collecting books, he carefully filed the names and mementos of his exile friends who had passed away. For the aging exiles, a funeral is not only an opportunity to let go of and mourn their comrades but also a meeting place for the Indonesian exile diaspora in Europe. Basuki Resobowo – an exile and prominent painter – once painted a picture of a funeral that included a sign saying *Perkumpulan Kematian Indonesia* (Indonesia Association of Death) (Setiawan, 2010), an ironic abbreviation of PKI and an obvious depiction of the exiles' condition.

The scholar–exiles have suffered increasing uncertainty and vulnerability throughout their post-1965 journeys, and the sudden change from student to exile changed what at first was regarded as political precarity into the precarity of their "overall existence" (Casas-Cortés, 2014). However, the precarity of the exiles is not limited to their uncertainty about life direction and their constrained work conditions: it is also about their existence and endeavors as exiles, their forever connection in disconnection with the homeland, the state regime that they despise but can do next to nothing to change, and in being trapped between the yearning and hesitancy to detach.

Through different activities, scholar–exiles live with and, one may say, make use of their precarity to achieve their activist goals. Many of them are involved in the exiles' solidarity collectives. Several social groups, organizations, study groups, and foundations have been set up by the exiles to connect with each other. The groups are diverse in nature; some are simple social groups based on geographical location, while others are used to mobilize their social and political agency. Many exiles joined or established socio-politico-human rights organizations, such as Tapol – which campaigns for the rights of 1965 ex-prisoners in Indonesia – and LPK65 or the 1965 Victim Defender Institution.

In addition, each exile has their personal way of creatively dealing with pain and precarity. Melanie started a free medical practice for undocumented Indonesian workers in the Netherlands. Tom became an advisor to a leftist youth political party in Indonesia while frequently hosting Indonesian students at his home in Sweden and receiving personal visits from Indonesian activists/scholars. Kuslan wrote poems and essays, engaged in scholarly and public discussions both offline and online, occasionally hosted Indonesian students, and once joined the art community group in his neighborhood in Woerden.

The exiles also tried to turn their absence into a presence by connecting with the Indonesians at home. Since the early 2000s, the Internet has provided a useful platform for some exiles to share their ideas and knowledge via mailing lists and blogs.

These active engagements have gained more exposure and have intensified since the fall of the Suharto regime. Many more scholars in exile express what they think and feel, telling their life stories, channeling their psychological baggage, and writing about their sense of loss due to being exiles. The topics and forms of their writings, which are mostly in Indonesian, are diverse, ranging from poems to memoirs, from political opinions to love stories, and from historical accounts to contemporary commentaries. In these conscious efforts to reconnect themselves to the home country through their writings, they produce and share alternative knowledge derived from their ideas, knowledge, and experiences as scholars in exile.

Chambert-Loir (2016) listed all the works of the *eksil* people in *sastra eksil* (literature of the exiles), referring to it specifically as "the writings of Indonesian authors constrained to live in foreign countries for political reasons after the putsch of September 30, 1965" (p. 119). He compiled a list of 133 volumes, namely, 37 collections of essays, 30 books of poetry, 3 plays, 6 collections of short stories, 15 novels, and 42 autobiographies. The numbers are astounding compared with the 25 memoirs written by the group of 1965 political prisoners imprisoned on Buru Island, who were politically exiled inside the country until the end of the 1970s. It shows that Indonesia lost a large number of potential intellectuals as a result of 1965 politics (Chambert-Loir, 2016).

The precarity and the different responses of each exile are shown in *sastra eksil*. Their ambiguity and in-betweenness are also depicted. Their longing for and fascination with the home country are expressed, as are the pain and failure of being political exiles, which led to resentment and bitterness toward the regimes. They are caught between the nostalgic remembrance of friends, family, and past relationships, and the sense of alienation and fear for relatives, of sufferings and disappointments, on the one hand, and hope and the energy to vocalize their existence, on the other.

Through these written works, specifically the life narratives, the exiles have materialized their personal engagement with precarity – which Chambert-Loir (2016) rightfully referred to as "a rationale in lives shattered by the events of 1965–1966 and that consequently went through rare turbulence independently of the individuals' control." The narratives, he argued, are stories of repeated failures but contain the moral stories that "life goes on, and for some, the struggle goes on too" (p. 216).

Dragojlovic (2012) argued that the exiles' insistence on failure is not only intended to provide knowledge about themselves and the Indonesian left, but also to incite a discourse on the possibility of repeating the potential loss of the past in the future. By incorporating their narratives into mainstream narratives, they actively rework the absence, turning it into an active presence and initiating a dialogue between past and present. The narratives of the 1965 exiles have never been given a place in the official history book or in the state archives. Their experiences, knowledge, and ideas became "the imagined" history less known by many Indonesians, at least not until the fall of the Suharto regime in 1998. Their initiative to share their narratives is an effort to fill the gaps in Indonesian history about the 1965 period.

Yet, to realize those purposes, their narratives must reach the desired audiences, namely, the broader Indonesian people and especially the post-1965 generations. The difficulty in communicating to the desired audiences may cause the narratives to become more abstract and hollower to those in their home country (Sreberny-Mohammadi & Mohammadi, 1987). Moreover, with the continuing state censorship of "communism-related" books, their writings are not widely distributed in Indonesia.

Intersecting student mobility trajectories across time and space

The exiled Indonesian scholars have other ways of communicating their narratives. Informal and formal group discussions, personal meetings, scholarly discussions, and other activities with Indonesian students are frequently organized by several exiles in Europe. Through direct interactions with the young generations of Indonesian international students, elder scholars in exile disseminate their alternative, suppressed, and silenced version of Indonesian history. They do so because they feel obliged to fill the students' knowledge gap. Today's Indonesian students are a generation whose memories and knowledge of 1965 are dominated by the national/political narratives of Indonesia's New Order regime. Hirsch (2008) used the term "post-memory" to describe the intergenerational and transgenerational transmission of traumatic knowledge and experience that has such a deep effect that it feels as though it is the person's own memory. The Indonesian post-1965 generation resembles the post-memory generation, but their 1965 memory is instead filled with and shaped by the trauma and fear narratives of communism planted by the state regime.

The connection between the current students and the exiles is a result of the latter's mobility trajectory coincidentally crossing the mobility trajectory of the students who study near where the exiles live. The two generations come into contact in various ways, ranging from group activities, discussions, and informal meetings to personal contacts. In the Netherlands, in particular, where most of the 1965 exiles live and many young Indonesians study, there has been much interaction between the exiles and today's Indonesian students.

The connection between the exiles and the students is a process of getting to know each other and interacting. Tria, who completed a master's degree in development studies, recalled the live interaction she experienced during her first meeting with exiles. She was amazed by the stories shared by seven of them, but she was also made to reflect on her own motive and stance when they asked about her interest in the 1965 issue. The exiles generally ask questions about interest and purpose when they first meet the younger generations. It is an automatic curiosity, as the exiles know that they are depicted as communists in the historical narratives of the New Order regime. It also shows how careful and reserved they are before opening up to others. The detailed questions do, however, provide a clue as to how the young students recognize the exiles' trauma and precarity. As explained by Tori, a postgraduate student who regularly interacted with the exiles when he was studying in Leiden:

> They usually asked: Who are you? Why do you want to know? What are you studying here? It took me some years to get to know them personally. I knew that, for example, Mr. Wardjo was afraid of meeting any Indonesians during his first years here. Fear of being persecuted still haunts them, and so those investigative questions are quite understandable.

As a leftist activist, prior to his study, Tori had heard about 1965 and the exiles. Even so, the direct interaction with the exiles gave him a greater understanding and more knowledge of his country's history. Having listened to their stories and observed their responses and reactions to various subjects, he was able to summarize the situation as follows:

> Exiles are often not regarded as victims because they live abroad. I came across many issues that I had never imagined. They experienced such trauma that they are often paranoid, suspicious, and closed. Although they might tell stories about 1965 to students they don't know personally, they would not make any mention of themselves when doing so. . . . There are many aspects of the 1965 tragedy that we have to bear in mind. For me, the exiles are like a bridge to the past . . . a continuity between the present and the past.

The exiles' historical account of what happened in 1965 is just one way the young students learn about the exiles' precarious and traumatic experience caused by 1965 politics. For students who know very little about the exiles, listening to their stories is fascinating and causes them to reflect on their own situation as students. Tria admitted that she had a mix of contrasting emotions when meeting the exiles:

> I was ashamed because, as a member of a younger generation, I was less eager to follow the social and political development of the country than the grandpas and grandmas that have been living in the foreign country for decades. Many of us, the younger generation, have not yet spent even one year in a foreign country, but we are already out of touch with the social and political changes back home. But I also feel proud to have the opportunity to get to know them more closely, to listen to their travel stories.

The similarity of their position as Indonesians studying abroad also prompted empathic reflection. Bima, who did conflict studies in The Hague, reflected on his interaction with the exiles from his positionality as a student:

> The thing that affects me the most is that the exiles were students sent and prepared as agents of development in the Sukarno era, and how the dispute in Indonesia impacted the students abroad. As an international student, I could be in the same situation and face the same consequences as them, being an exile, if there is a fundamental change in Indonesia.

Another example of their reflection is based on racial positionality, as experienced by Tania, who works as a gender activist and is doing a master's degree in gender studies in The Hague. She talked about the things she had learned from her personal visits to Melanie, a 1965 exiled student of Chinese descent:

> I felt I had been fooled by the Suharto regime. I learned and was enlightened. I learned about the impact of 1965 on the Chinese-Indonesians. Also, the Chinese-Indonesian exiles' stories convinced me that the Chinese-Indonesians are also fighters and Indonesians.

As Tania realized, Melania's story of experiencing exile helped her learn about a fascinating nationalist struggle that she had not expected to learn about Chinese-Indonesians. This is in contrast to the dominant narratives inherited from the New Order regime, which frames Chinese-Indonesians as economic subjects whose loyalty to Indonesia is in question.

The nationalism of the exiles is a prominent aspect that was often mentioned by current students. Most of them admire the exiles' sense of belonging and their bond with Indonesia, even after displacement by the political regime. This is demonstrated not only by the stories shared by the exiles, but also by their actions and engagement in Indonesian issues. Soraya, who studied law and now works as a migrant worker activist, explained the experiences as follows:

> I was involved with the exiles in fighting for the rights of Indonesian migrant workers. The exiles gave their support by providing a place for the meeting, participating in raising the issues of migrant workers et cetera. The exiles are the inspiration for continuing to fight for justice.

The exiles' views on Indonesia are critical in many ways. In addition to their subscribed ideology, their precarious experiences, series of disappointments, and unfulfilled expectations also contribute to their views. Furthermore, there is a spatial and temporal distance in viewing Indonesia. Despite following the country's development closely, the Indonesia they knew has changed dramatically and they are not as familiar with it as in the past. For some students, the spatial and temporal distance creates obvious differences in their views on Indonesia. As Bima put it, "The Indonesia the exiles lived in is a different Indonesia in time and locality."

The interaction with the exiles also creates an understanding that the memory of 1965 is viewed differently even among the victims. The young students learn that the exiled scholars are not a homogenous group and have different views. The differences that divide them are not limited to how they view socialist and communist ideology, but include their diverse opinions about the role of the PKI delegation in handling its members (including students abroad) during the political crisis in Indonesia. Furthermore, there are also exiled scholars who do not wish to be associated with particular, dichotomous opinion groups such as pro- or contra-Sukarno, pro- or contra-delegation, etc. Tori normalized the division of exiles as just "various kinds of socialists." Interaction with various kinds of exiles made him realize that they cannot be categorized simply as communist or leftist, but that they consist of various groups that are often in conflict. These encounters also help younger generations realize that the exiles are far from a harmonious single entity. The recognition of such diversity is important in unsettling the common and unquestioned perspective on 1965, which represents the victims as homogenous.

In sum, the new generation of students learn new memories and new knowledge when their path intersects that of the exiles. The memories and knowledge are not directly internalized but are processed through their own backgrounds, positionalities, and reflexivities. Many view the exiles' precarity as a painful result of the 1965 tragedy, but it also encourages them to fight for and empathize with this marginalized group. However, some are still affected by the trauma narrative that lingers along with the fear of communism in the country.

Conclusion

The lived experiences of the exiled scholars in this chapter illustrate the uncertainty and vulnerability created by their sudden transformation in the 1960s from privileged students into precarious exiles. Although not all exiled scholars engage by being politically active, reaching out to students and revealing their life stories and opinions on issues in modern Indonesia are also acts of political stance and defiance. In doing so, they counter the single national narrative that is hegemonic in the minds of today's Indonesian youth. This shows the expanding effect of the exiles (Roniger, 2016) that is driven by their precarity.

It is also important to note that this highlights the intersecting of student mobility trajectories. How the trajectories of today's students cross those of the exiles and have an impact on the former's trajectories illustrates the temporal dynamics of student mobility. This temporal stretch expands the tendency to view the mobility trajectory as a single temporality of an individual. Beyond the exilic context and within the broader international student mobility discussion, the overlapping student trajectories underline the possibility of the cross-generational impact of exilic mobility.

Acknowledgements

This chapter draws on a broader research programme supported by The Dutch Research Council (NWO) (Aspasia Prize for Maggi Leung, awarded in 2012). We are grateful for the interviewees' invaluable stories and trust in us.

References

Casas-Cortés, M. (2014). A genealogy of precarity: A toolbox for rearticulating fragmented social realities in and out of the workplace. *Rethinking Marxism, 26*(2), 206–26. doi: 10.1080/08935696.2014.888849

Chambert-Loir, H. (2016). Locked out: Literature of the Indonesian exiles post-1965. *Open Edition Journal, 91*, 119–145.

Cornejo, M. (2008). Political exile and the construction of identity: A life stories approach. *Journal of Community & Applied Social Psychology, 18*(4), 333–348. doi: 10.1002/casp.929

Dragojlovic, A. (2012). Materiality, loss and redemptive hope in the Indonesian leftist diaspora. *Indonesia and the Malay World, 40*(117), 160–174. doi: 10.1080/13639811.2012.683670

Eickhoff, M., van Klinken G., & Robinson, G. (2017). 1965 Today: Living with the Indonesian Massacres. *Journal of genocide research, 19*(4), 449–464. doi: 10.1080/14623528.2017.1393931

George, T (2016). Precarity, power and democracy. In N. Buxton & D. Eade (Eds.), *State of Power 2016*, (pp. 131–146). Amsterdam: The Transnational Institute. Retrieved from https://www.tni.org/en/publication/precarity-power-and-democracy

Gurning, A. T. (2011). *Memory, experience and identity of the Indonesian political exiles of 1965 in the Netherlands* (Unpublished Master's thesis).

Hearman, V. (2010). The last men in Havana: Indonesian exiles in Cuba. *RIMA: Review of Indonesian and Malaysian Affairs, 44*(1), 83.

Hill, D. T. (2010). Indonesia's exiled left as the Cold War thaws. *RIMA: Review of Indonesian and Malaysian Affairs, 44*(1), 21.

Hill, D. T. (2014). Indonesian political exiles in the USSR. *Critical Asian Studies, 46*(4), 621–648. doi: 10.1080/14672715.2014.960710

Hill, D. T. (2020). Cold War Polarization, Delegated Party Authority, and Diminishing Exilic Options: The Dilemma of Indonesian Political Exiles in China after 1965. *Bijdragen tot de taal-, land-en volkenkunde/Journal of the Humanities and Social Sciences of Southeast Asia, 176*(2–3), 338–372.

Hirsch, M. (2008). The generation of postmemory. *Poetics Today, 29*(1), 103–128. doi: 10.2307/2754184

Hsiau, A. C. (2010). A "generation in-itself": authoritarian rule, exilic mentality, and the postwar generation of intellectuals in 1960s Taiwan. *The Sixties: A Journal of History, Politics and Culture, 3*(1), 1–31.

Lebang, T. (2010). *Sahabat Lama, Era Baru: 60 Tahun Pasang Surut Hubungan Indonesia-Rusia [Old friend, new era: Sixty years of ups and downs in the Russian-Indonesian Relationship].* Jakarta: Grasindo.

Ma, S.-Y. (1993). The Exit, Voice, and Struggle to Return of Chinese Political Exiles. *Pacific Affairs, 66*(3), 368–385. doi: 10.2307/2759616

Mudzakkir, A. (2015). Living in exile: The Indonesian political victims in the Netherlands. *Jurnal Masyarakat Dan Budaya, 17*(2), 171–184. http://jmb.lipi.go.id/index.php/jmb/article/view/282

Nadzir, I. (2018). Reclaiming Indonesian-ness: Online-offline engagement of Indonesian Exiles in the Netherlands. *Masyarakat Indonesia, 44*(1), 15–30.

Robinson, G. (2017). "Down to the Very Roots": The Indonesian Army's Role in the Mass Killings of 1965–66. *Journal of Genocide Research, 19*(4), 465–486. doi: 10.1080/14623528.2017.1393935

Roniger, L. (2016). Displacement and testimony: Recent history and the study of exile and post-exile. *International Journal of Politics, Culture, and Society, 29*(2), 111-133. doi: 10.1007/s10767-015-9201-7

Roniger, L. (2017). Citizen-victims and masters of their own destiny: Political exiles and their national and transnational impact. *Middle Atlantic Review of Latin American Studies, 1*(1). https://static1.squarespace.com/static/57f58 8b8cd0f68cfa768c7dd/t/58cb06of2e69cf51cc49e31e/1489700374868/2-MAR LAS-Luis+Roniger-Citizens-Victims-final.pdf

Said, E.W. (2002). *Reflections on Exile and Other Essays.* Cambridge: Harvard University Press.

Setiawan, H. (2010). Some thoughts on Indonesian exilic literature. *RIMA: Review of Indonesian and Malaysian Affairs, 44*(1), 9.

Sipayung, B. A. (2011). *Exiled memories* (MA thesis). Institute of Social Studies Erasmus University of Rotterdam, The Hague.

Sreberny-Mohammadi, A., & Mohammadi, A. (1987). Post-revolutionary Iranian exiles: A study in impotence. *Third World Quarterly, 9*(1), 108–129. doi: 10.1080/01436598708419964

Wright, T. C., & Oñate Zúñiga, R. (2007). Chilean political exile. *Latin American Perspectives, 34*(4), 31–49. doi: 10.1177/0094582X07302902

The politics of forced internationalization

The moral economies of "research in exile"
Rethinking a field from the perspective of reflexive migration research

Isabella Löhr

In the wake of the violent conflicts in the Middle East many scholars sought to leave their countries because they suffered from censorship, were dismissed, prohibited from working or they faced immediate physical danger to their lives. Subsequently, the political persecution of scholars developed into an issue that gained attention from academic circles in Europe and North America. A large number of private and public, local, national, and European initiatives support scholars and students who seek asylum, live in refugee camps, or are in urgent need of support. Joining forces with long-existing initiatives such as the British Council for At-Risk Academics (CARA) or the two New York based Scholar Rescue Fund (SRF) and the Scholars at Risk Network (SAR), they pursue several aims at a time: to protect the lives of scholars and their families, to take a public stand for academic freedom and to enable what is presumed to be the core of higher education, the sharing and transfer of ideas, arguments and paradigms.

What can be hailed as substantial attempt to defend individuals and the core values of modern science, has also its drawbacks. First, to predicate fiercely desired entry permits to escape persecution upon the evaluation of professional performance by scientific boards presents serious ethical dilemmas. Second, humanitarian motives and the vision to enhance some sort of science dedicated to peace and a universalist dialogue should acknowledge that programs for threatened scholars are historically, socially, and culturally situated and that they participate in the governance of migration – that is, an all-encompassing management system that controls the movement of people by setting up laws, regulations, measures, and procedures that advantage the movement of one group – for example, the "highly skilled" – to the detriment of other groups – such as "refugees" – whose movement is hindered by complex border control mechanisms. Third, advocating the cause of scholars at risk within the political and moral framework of existing migration policies means carrying the political imbalances of current migration regimes into the university and consequently entangling the politics of migration with prevalent knowledge orders and practices in the university.

It is the contention of this chapter that scholars and university activists need to think about the epistemic and moral frames that inform the drafting of support programs and their implementation as well as the way they speak about and produce knowledge on scholars at risk. More specifically, the chapter posits that help schemes for scholars at risk tend to contribute to the reproduction of policy categories that present migration as external to modern societies and as a problem for social cohesion and the governance of societies. Moreover, the chapter argues that most programs suffer from a conflict of values that stems from divergent moral economies that are inherent in the structural setup of the programs. Following the debate on the so-called reflexive turn in migration studies, the chapter suggests that helping scholars at risk and the way in which individuals, institutions, and private initiatives make sense of this needs to be written back into the field of migration studies and its recent attempts to investigate the categories used to understand and deal with (forced) migration and different forms of mobility. To support this assertion, the chapter denaturalizes the logics that inform the support schemes for scholars at risk. It argues that initiatives to balance forced migration in the university sector contribute to the political-legal categorizations of ethnic and social groups. In doing so, they feed into the selective politics of migration control that enable the movement of particular groups, while EUropean borders and human rights mechanisms remain closed or unattainable for the majority of displaced people. Moreover, the chapter shifts attention to the conflicting logics and values that inform "research in exile" and that result from the fact that the field connects very different contexts with one another – the principles of migration governance, knowledge practices, and university-led activism. It emphasizes the need to make those conflicting values visible and to open up room for a self-conscious discussion.

I do not intend to question any initiative for persecuted scholars and draft measures to remove academics from life-threatening situations. My concern is a different one: I want to sensitize to the fact that the terms, concepts, and methods used to act and speak in favor of scholars at risk are not self-evident or uncontested but form part of the political and moral governance of migration. Consequently, it is necessary to reflect upon the extent to which current programs enshrine, refashion, or disrupt previously established migration-related configurations and how this becomes incorporated into the discursive knowledge on and management of the individual and professional prospects of the scholars at risk. I argue that only a reflexive perspective on the positionality of "scholar rescue" in the field of migration governance and the rootedness of its language and concepts in the discursive knowledge on migration allows us to understand the imbalances, limits, and opportunities of this sort of activism.

The chapter consists of three parts. The first part presents the main arguments from reflexive migration research. It carves out the difference between policy-oriented migration research and knowledge production that develops its research agenda on the basis of a continuous self-interrogation, which discloses its own social position and reflects its moral premises and values. Part two situates academic support schemes at the "nexus of academia and activism" (Kasparek & Speer, 2013, p. 259), and it elaborates on the conflicting rationalities and values that result therefrom. Taking the German Philipp Schwartz Initiative as a case study, the section argues that support for threatened scholars is embedded in a particular knowledge order that only becomes visible if we perceive programs in support of threatened scholars not as a self-evident and universal endeavor but as a historically, spatially, and institutionally situated practice that is inscribed in colonial legacies of modern science and built upon specific professional value systems and expectations. The third part examines the connections between "research in exile" and migration politics. Starting out from current discussions in migration research on borders, mobilities, and migration regimes, it analyzes the epistemic and moral frames that have guided the governance of academic mobility since the early twentieth century. It argues that the perception and management of migration in higher education is closely linked with the political categorization of different forms of mobility.

Reflexive migration research and the moral economies of knowledge production

Migration research is witnessing a "migration knowledge hype" (Braun et al., 2018, p. 9) that was initially situated in the context of the events in

2015 when thousands of people arrived in the European Union. This has led to an increasing demand for expert knowledge in politics, media, and the public. Subsequently, migration experts were expected to create "useful" knowledge that would provide the basis for drawing up evidence-based, more restrictive policies to regulate and control migration movements from the Middle East and North Africa (MENA) region to the EU (Scheel & Ustek Spilda, 2019). While this led to a remarkable growth of funding schemes and the foundation of migration-related research institutes, it also prompted strong criticism from within the interdisciplinary field of migration studies that problematized the impact of the, at times, close relationship between research and policymaking (Stierl, 2020). Though this perspective has become more prominent, it is not new. As early as the 1990s and 2000s, social scientists pointed out that many studies on migration, mobility, and forced migration perpetuate nationalized notions of society that take concepts of national territory, belonging, and non-belonging for granted (Malkki, 1995). Similarly, scholars from refugee and forced migration studies discussed the problems of research designs that privilege the world views of policymakers and practitioners by taking their "categories, concepts and priorities . . . as their initial frame of reference" (Bakewell, 2008, p. 432). This has cemented policy labels and produced figures to whom a particular, mostly reductive set of rights and options is assigned (Zetter, 2007).

More recently, the reflexive engagement with epistemic frames and the drafting of one's own practices of categorization has been transferred to other migration-related subjects. For example, Rogers Brubaker (2013) critically reflected on the study of the religiously defined immigration population in European countries. Taking the category "Muslim" as a starting point, he argued that scholars actively contribute to establishing epistemic frames that define a particular social group in religious terms with the result that alternative sociopolitical frames of reference to explain, for example, marginality recede into the background. Consequently, he stressed that scholars must differentiate between politics and analysis by making categories that are politically contested to their object of analysis instead of taking them for granted. In a similar way, recent analysis focuses on the paradigm in Western Europe to integrate different ethnic groups and provide moral monitoring of their efforts. Schinkel (2017) and others critically argued that this integrationist focus often construes a specific image of society as a homogenous, stable entity and an essentialist understanding of ethnic and cultural differences, with certain groups assumed to be in particular need of integration.

Accordingly, an increasing number of scholars have called for a more self-conscious perspective on the production of knowledge on migration. They have proposed to conceive migration-related categories as a means to produce differences because of ethnicity, class, gender, or religion and to direct attention to the epistemological underpinnings of their writings. The vision is to embark upon a research that disconnects the field from the migration apparatus and perceives movement as an integral part of society and social theory (Bojadžijev & Römhild, 2014; Dahinden, 2016). A reflexive perspective perceives migration not as a self-evident, given object of analysis but as a product of changing constellations and categorizations that are themselves used to allocate resources and reorder sociopolitical hierarchies. It asks how research itself contributes to constituting migration as a "social fact" (Löhr & Reinecke, 2020). Drawing on multiple sources – science and technology studies, feminist standpoint theory, postcolonialism, methodological nationalism, transnational approaches, or social constructivism – the different strands of reflexive migration research join together in an attempt to investigate the categories used to make sense of and deal with migration and different mobilities. The challenge is to acknowledge that research on migration always forms part of the social negotiation of what a migration-related phenomenon is.

Part of this endeavor is to understand which moral premises migration research or, more generally, the production of knowledge on migration has been built upon. In this context, Boris Nieswand (2021) proposed to position the knowledge production on migration in a moral framework that comprises political discourses, expert knowledge, and commonplace moral judgments alike. He argued that because migration is socially and politically contested, researchers always take a particular stand in this debate. Though he conceded that concepts of good and bad inherent in the knowledge that social sciences produce differ from other techniques of moralization, he posited that their underlying logics and effects are nonetheless comparable as they hierarchize persons and groups. A reflexive perspective, Nieswand (2021) concluded, allows for a meta-perspective that reflects upon the opportunities and limits of scientific knowledge production. It scrutinizes the social position of researchers, their relationships to their objects of research, and the leading questions, assumptions, or paradigms.

In the following, I seek to show that we can fruitfully transfer this debate to the field of "research in exile". Though knowledge production is mostly implicit, the fact that it is deeply rooted in the university allows us to question the field about the relationship between humanitarian and scientific values and about its relationship to the contested governance of migration. "Research in exile" forges a particular relationship between knowledge and forced migration that condenses university, activism, knowledge production, and migration governance. Imbued with moral concepts of mobility and knowledge production, it assigns forced migration a specific position in the university, and it informs the discursive knowledge on the relationship between university and forced migration. Following Didier Fassin (2009), the term *moral economies* is used to direct attention to the political dimension as well as to the complexities and contradictions in which "research in exile" is embedded. The subsequent sections make these complexities and contradictions visible and open up the space for a reflexive perspective on "research in exile".

The double grounds of research and activism: Humanitarianism, employment prospects, and postcolonial legacies

In the following, the focal point is how the link between humanitarian assistance and knowledge practices is framed within the university. I will explore the intersection of the university, activism, and knowledge and elaborate on the conflicting rationalities and values that result therefrom. More specifically, the section argues that support for threatened scholars is embedded in a particular knowledge order that only becomes visible if we perceive "research in exile" not as a self-evident and universal endeavor but as a historically, spatially, and institutionally situated practice that is inscribed in colonial legacies of modern science and built upon specific professional value systems and expectations.

The connection between activism and research is not new. A number of research fields in the social sciences deal with politically and socially contested issues, with some researchers aiming to intervene in public debates and policymaking. Prominent examples are feminist research and gender studies that critically engage with power dynamics and patterns of discrimination, or critical border studies that intend to denounce and work against the consequences of the border enforcement mechanisms at the EU border.

Seeking the practical implications of their findings, scholars problematize and visualize unequal processes of integration or exclusion, the uneven distribution of access to resources, or the discrimination of groups because of language, religion, gender, education, social background, citizenship, or ethnicity-related classification. To achieve this, they produce empirical material in order to challenge official reports, statistics, or maps (Casas Cortes & Cobarrubias, 2018), or they explore new ways of narrating migration-related discrimination (Wonders & Jones, 2019). Most of this research develops within the framework of universities. They provide the institutional and infrastructural environment required, such as the material prerequisites (salaried positions, libraries, databases, and research funds) and an affiliation that transforms the researchers into full and acknowledged members of the scientific community with evidence-based research methods and evaluation procedures as core elements. Ultimately, their activities profit from their position as researchers who are free to think, a position which is guaranteed by the universities' constitution.

University-led refugee initiatives also sit at the crossroads of research and activism. They share a number of common features with activist research (Carstensen et al., 2014), such as the interventionist character, the activation of the people concerned, and writing as a means to garner support, to reflect experiences, or to map the field (EUA, 2020; Watenpaugh et al., 2014). At the same time, the relationship among the university, knowledge practices, and activism is different, as weights are shifted from knowledge production to assistance for individuals. Support schemes intend to enable threatened researchers to resume their research activities by relocating them to an institutional context that provides for physical, intellectual, and academic integrity. To this end, universities and research institutes work in close cooperation with non-governmental organizations such as the SAR network or CARA. They gather information on the overall political situation in the respective countries, they single out the researchers who are in immediate danger or urgently need support, and they match the scholars with universities. However, we would not fully grasp the complexity of the mechanisms at play if we were to reduce humanitarian assistance for scholars solely to the practical measures to help people in need. The support schemes are firmly rooted in the university, they build upon the rules and practices of the scientific community, and they are inscribed in epistemic frames about what makes up "good" science. Thus, it is imperative to determine how these connections become manifest in the support schemes.

The German Philipp Schwartz Initiative may serve as a case in point. The Initiative grants funds to research institutes or universities that want to host a scholar, and it is part of the Alexander von Humboldt Foundation (AvH), a government-funded research foundation that advances the internationalization of German higher education by sending researchers abroad and by offering international scholars research stays in Germany. AvH's selection committees consist of renowned members of the scientific community, in the case of the Philipp Schwartz Initiative, including the German National Academy of Science – Leopoldina and the German Research Foundation. It is fair to say that the initiative attempts to square the circle in that it is situated at the juncture of the moral economy of humanitarian assistance and an imperative to achieve scientific excellence, yet with a clear tendency to give preference to the professional prospects. Though the program intends to help "researchers who are demonstrably at risk" (AvH, n.d.), the initiative aims at transforming the emergency situation into something that ultimately serves scientific knowledge production. The funding criteria bring scientific eligibility and prospects to the fore. Applicants are not the scholars but the university or research institute that wants to host a scholar because of the researcher's track record. Consequently, the granting of the sometimes lifesaving scholarships is bound to a peer review process that decides the "scientific qualification of the guest" (AvH, n.d.) and evaluates if the professional profile matches that of the host institution. Against this background, career prospects play a crucial role. Scholars have to "have the potential to be integrated in the (broader academic) employment market," a precondition that is mirrored in the requirement that candidates must have completed a doctorate in medicine or law and are obliged to "provide evidence of equivalence to a research PhD" (AvH, n.d.). The Philipp Schwartz Initiative is firmly embedded in the research apparatus of universities. The imperative to help is situated in an environment that is structured by scientific standards and routines of the host country, which translates the humanitarian situation into a professionally framed set of skills and expectations. The field thus intertwines two perspectives on scholars at risk that are hardly dissoluble: individual researchers in need who receive help under the condition that they create a research-related added value and a structural setup that is situated at the interface of humanitarian assistance, migration policy, and knowledge production.

Amazingly, this setup remains silent in two regards. First, it does not attempt to connect with ongoing debates on how to put Dipesh Chakrabarty's

(2008) well-known claim to provincialize European thought into practice. Since the 2000s, the social sciences and humanities, history in particular, have tackled the epistemic legacies of today's knowledge production and politics. Connecting these with the imperial and colonial past, they assumed that the asymmetrical power relations between European and non-European countries since the nineteenth century were at the heart of the formation of modern science as the dominant institution of knowledge production in the modern world. This perspective inscribes the history and sociology of science and knowledge in its colonial past. It points out that the colonization of large parts of the world was to a large extent based on the process of acquiring and institutionalizing botanic, geographic, medical, demographic, legal, or ethnographic knowledge about the populations and countries incorporated into colonial rule (Bauche, 2017). A postcolonial critique of academic epistemologies contested the universal claim of basic concepts, and more importantly, the very process and strategies with which the production of knowledge is legitimized and presented as evidence-based and, accordingly, purportedly universal knowledge (Sivasundaram, 2010). This also holds true for universities and research institutes that situate and nourish these bodies of knowledge for the knowledge practice of academic disciplines and for the strategies and mechanisms with which they reproduce their claim to a knowledge monopoly. Against this background, Shalini Randeria and Regina Römhild (2013) emphasized that the non-reflection of the colonial past of science and knowledge results in a Eurocentrism that has European (and North American) knowledge production and the respective university values at its unnamed center. Even though the transfer of these values and institutional schemes has produced a wide array of knowledge practices worldwide (Shils & Roberts, 2004), the very fact that European and North American universities possess a significant institutional, financial, and discursive power reinforces their hegemonic positions and the very asymmetries that enabled the rise of European concepts of science and university to a global standard.

This postcolonial critique of science and knowledge production mirrors the mechanism at play in the process of selecting scholars at risk for fellowships that enable them to leave threatening constellations. At the core of the selection procedure is the question of whether the applicants meet with specific standards of knowledge production, yet without historicizing these standards and situating them in their cultural, political, and social contexts. As a result, they are presented as being given and purportedly universal instead of perceiving them as the historical outcome of the dynamic entanglement between Europe and its former colonies. We must take this consideration into account for two reasons. First, most of the scholars at risk come from regions that were formerly under colonial rule, and second, this has left its traces in the building of the respective systems of higher education, which in most cases are not competitive with the universities in the countries of application. Thus, the selection criteria reinforce the predominance of Western standards in the field of science and knowledge to the detriment of the scholars at risk.

Moreover, the programs have remained silent about what Asli Vatansever (2018, p. 153) called "academic work under precarious conditions" (see also Vatansever in this volume). She highlighted the inherent contradictions in current support schemes that simultaneously involve a humanitarian discourse that victimizes the scholars, their empowerment by offering application opportunities, and the limits of a highly competitive academic labor market that produces precarious subjects, which become even more precarious with the missing prospects of a steady position in the context of a "forced nomadic way of living." While Vatansever has striven to turn the structural and political precariousness into a resource that might challenge established and institutionalized forms of intellectual subjectivity, her argument made another point visible that has been left unsaid. As I have argued elsewhere (2021), we can only understand the patterns of the forced mobility of scholars if we write it back into the global history of academic mobility and conceive of it as labor migration. From this perspective, we must then ask how the support for scholars at risk affects or even co-produces the categorization of other forms of mobility.

Border politics: Higher education and the governance of migration

This section expands on the intersections between "scholar rescue" and the governance of migration and carves out how the endeavors for researchers in distress are connected with political discourses and measures. To this end, it seeks to uncover the epistemic and moral frames that "research in exile" shares with restrictive migration policies. In so doing, it highlights the dynamic interplay between the embodied individual mobility of scholars and the selective governance of migration. The section gives a brief overview of recent findings on borders, migration regimes, and mobility, and it suggests that they provide the analytical framework to contextualize the mobility options in higher education. The section seeks to create awareness of the fact that support for threatened scholars is always inscribed in the contingent and contested politics of migration.

Mobility studies have suggested rereading the emergence of modern societies through a mobility lens and conceiving of mobility as an "ensemble of highly meaningful social practices that make up social, cultural and political life" (Adey et al., 2014, p. 3). Mobility studies have emphasized the relational character of mobilities in at least two respects. First, they propose to connect different forms of mobility and to ask how different categorizations of movements interrelate with one another and how specific forms of mobility are framed in political, legal, or sociocultural terms. Second, mobility studies have proposed to think of mobility and immobility as relational concepts that constitute each other. The invitation to question mobility categories instead of taking them for granted has brought the question to the fore as to how the interplay among categories of (im)mobility contributes to drawing the line between inside and outside, belonging and non-belonging, and exclusion and inclusion (Sheller, 2018). Border studies add to this perspective a critical engagement with the normalization of national borders and the political rationalities and value orders that inform border regimes (Hess & Schmidt-Sebdner, 2021). They conceive of geopolitical borders as a complex assemblage of control and contestation that is shaped by a multitude of divergent actors, forms of agency, knowledge formations, and sociopolitical practices. Border studies point to the multi-locality of borders, or, in the words of Etienne Balibar (2002, p. 84), to "the ubiquity of borders" with its overall rationality to control and prevent (un)desired movement. From this perspective, border regimes perform multiple functions: they delineate the boundaries of national societies; they are the sites where ethnocentric visions of national societies are performed, implemented, and contested; they mobilize and immobilize; and, consequently, they are at the heart of the categorizing and governing of human mobility (De Genova & Peutz, 2010; Walters, 2015). Finally, research on migration regimes has contributed to this perspective a focus on the actual practices of control and mobility (Pott et al., 2018). This strand of research does not take migration as a starting point but rather the contested, complex, and at times contradictory nature of migration policies, which are perceived as a reaction against migrant practices. Thus, a regime perspective takes the agency of the migrants into account and connects it with "the intertwined patterns of global mobility and current orders of power and inequality" (Horvath et al., 2018, p. 302).

These approaches are of particular interest for our context. Borders, migration regimes, and the politics of (im)mobility are instruments with which "regimes of mobility" (Glick Schiller & Salazar, 2013) are established that create unequal social relations across the globe by doing both connecting and separating people. Thus, border and migration regimes differentiate between different forms of mobilities and allow for the selective channeling of people. A good example of this is the differentiation between illegalized migrants, the temporary movement of tourists, and the oftentimes encouraged movement of the highly skilled. These categories are connected with different sets of rights that enhance the movement of one group while immobilizing the other. Thus, migration politics produce the figure of the migrant, their, his or her (im-)mobilization and (il)legalization and, subsequently, the establishment of different classes of people (Mezzadra & Neilson, 2013).

This also applies to the university sector. The history of higher education illustrates that mobility options have always depended on how students or researchers were categorized and put in relation to other mobile groups. The history of Chinese migration to the US in the decades before and after the turn of the nineteenth century is well researched. We know a lot about how restrictive and racialized migration regimes contributed to constructing the figure of the illegal Chinese immigrant. Historians have elaborated on how the border regions of the United States were transformed into sites of contestation over illegal migration; immigration and labor policies; bilateral relations; and the framing of race, citizenship, class, and gender (Lee, 2007). However, less well-known is the travel of Chinese students to the US from 1908 onwards in the context of the so-called Boxer Indemnity Scholarships. This scholarship program was set up in 1908 in a bilateral treaty between the Chinese empire and the US and was intended to discharge some of the compensatory payments imposed on China by an international military alliance in 1900, in the aftermath of the violent repression of the Boxer Rebellion (Ye, 2001). The outstanding reparations were converted into scholarships with the aim of attracting the future Chinese elite to take up their studies in the US. In that way, the US government hoped to gain a dual benefit for its own global position. On the one hand, it wished to extend its sphere of influence in East Asia by influencing the far-reaching reform policies the Chinese government was introducing. On the other hand, it sought to enhance its own claim to ascendancy in the Pacific by supplanting the imperial capital of Japan as the main destination for Chinese students abroad with its own rising system of higher education. For this purpose, the restrictive immigration rules and quotas were rescinded for Chinese *students* while they were upheld for Chinese *migrants* (Hsu, 2015).

The politics of the selective categorization of human mobility continued and were adapted in the decades after the Great War, as the history of humanitarian assistance for researchers from Europe shows. The US-American immigration laws of 1921 and 1924 extended the restrictive immigration policies by means of quotas for specific regions of the world. But the laws stipulated exceptions for so-called "'nonquota' immigrants," a category that gave preference to immigrants with specific occupational skills. Professors and, after 1924, students were among those preferred groups if, in the case of professors, they had held an academic appointment for at least two years prior to their visa application (Lee, 2008, p. 14). From 1933 onwards, this stipulation allowed for the immigration of scholars who had lost their professional position and were persecuted because of denomination or ethnic categorization by Nazi rule. Although the US government extended its restrictive migration policy in the context of American isolationism, economic depression, and anti-Semitism to refuge issues (Zucker & Flink Zucker, 1996), persecuted scholars were exempted from the rule. They received entry permits provided that they had an employment contract with a US-American university before leaving the country (Krohn, 1993). Thus, the academic support schemes were built upon earlier classifications of migrant groups. They continued earlier politics of (im-)mobilization by exposing scholarly skills and emphasizing the practical value displaced scholars from Central Europe would bring to American universities. The underlying rationalities and values were comparable with the ones at the turn of the century: a competitive framework that perceived higher education as part of international relations and migration issues in particular as a means to underscore US-American claims to be a site of global knowledge production while access to most other sectors of the labor markets remained restricted on the basis of sociopolitical and racialized factors. Accordingly, programs in favor of scholars were eager to uphold the differentiation among refugees (connected with a number of fears and stereotypes, e.g., Communism, anti-Semitism, and labor competition) and the highly skilled. Their mobility options were maintained by detaching them from other migrant groups and focusing on their contribution to the advancement of American society.[1]

These patterns have been remarkably persistent over the course of the twentieth century, and they inform today's programs and initiatives. As the application rules of the Philipp Schwartz Initiative illustrate, the humanitarian side of being at risk plays a minor role when it comes to the drafting and implementation of university-related support measures. Currently, the same mechanisms apply as in the 1930s: while border procedures and opportunities to grant asylum for so-called third-country nationals are increasingly restricted on the European level (Wessels, 2021), the number of initiatives on behalf of persecuted scholars has steadily risen. In Germany alone, there exists an impressive spectrum of support structures, including networking opportunities (Academics in Solidarity [AiS], Chance for Science at Leipzig University, and the Off-University), institutional funding schemes at the federal (Philipp Schwartz Initiative) and state levels (e.g., the Hamburg Programme for Scholars at Risk (HPSAR), the *HessenFonds*, and the Einstein Foundation), and individual sponsorship opportunities (e.g., The Academy in Exile, the Scientific Integration Initiative). Moreover, the current schemes converge with those in the inter-war period with regard to the humanitarian aspect. Though the precarious humanitarian situation has triggered support measures, they are not enforced under the umbrella of human rights law but under that of labor legislation. The portal Science4Refugees of the European Commission puts it in a nutshell with the slogan "Welcome refugee researchers and students looking for a suitable job!" (EURAXESS, n.d.). This fits with recent findings that require us not to overemphasize the humanitarian but to take moralities, political motives, and contexts under close consideration. As Didier Fassin (2016) pointed out, moral grounds provide a powerful instrument for developing transnational, national, or local policies. For him, the practices of labeling mobile groups as refugees, asylum seekers, or refugee researchers involve more than a difference in vocabulary or status. Rather, they express a difference in the politics of recognition that emerges from the moral horizon of migration and asylum policies of a society at a given historical moment. What Fassin

---

1   Nonetheless, as Claus-Dieter Krohn (1993) has noted, labor competition and anti-Semitism were also contentious issues in the university sector, as well as the procedural obstacles that the US-American consulates in France and Portugal established regarding the visa issuance from 1938 onward.

(2009) called the moral economies of asylum policies entails not only the way in which societies speak about and create a problem-driven perception of mobility but also the far-reaching consequences for the groups concerned – their mobility options, their individual credibility, and the degree to which they are presented as society's other.

Conclusion

This chapter pursued a double goal: to examine the schemes of support for scholars at risk as a context-dependent endeavor that operates with specific notions of science, migration, and humanitarianism, and to trace the epistemic and moral values as well as the politics of science and migration that structure current efforts to assist threatened scholars. From this perspective, the patterns and rationales of support schemes appear in a different light, less optimistic, but as an integral part of the contested politics of migration and the historical past. However, a problem-oriented approach to "scholar rescue" does not question the field as such. Rather, it sensitizes us to the selective and uneven character of migration and refugee policies and helps us to comprehend its drawbacks. It discloses the epistemic and moral positions that imprint the drafting of specialized support schemes, and it furthers our understanding of the entrenchment of science and university in politics. This is what the analysis of the Philipp Schwartz Initiative and the connections of the field with European migration and border policies illuminate. They highlight the relational character of help schemes, their political enmeshment, and the complexities of the forced migration of scholars who move in and out of political and legal categories. Thus, a reflexive perspective allows us to better grasp the dynamics at play and to balance the humanitarian label carefully with what is at the core of "scholar rescue" – competitive and precarious academic labor policies and the contested negotiation of scientific standards and claims to universalism between scholars from the "Global South" and the "Global North."

In tracing these patterns, the chapter argues for a reflexive perspective on "scholar rescue" that displays the epistemic and moral positionalities and imbalances of higher education, as well as their contribution to the co-production of migration, and that scrutinizes the national order of higher education. Reflecting on "scholar rescue" can thus be a vantage point from which we can see the entire network of embodied, social, political, and academic relations that make up higher education. On the one hand, this allows us to grasp forced migration in academia as a social process in the course of which knowledge positions and transnational access to resources are negotiated. On the other hand, it becomes possible to search for the means by which we can decolonize the structural asymmetries between universities and knowledge production in different world regions and to provide for scholarly exchange on an equal basis prior to political threat. Such an approach places the agency and lived experience of the scholars and their capacity to act at the very center. Networking activities such as the Off-University, the AiS program, Chance for Science, the Academy in Exile, or Science4Refugees move in this direction. However, these need to be connected with a critical assessment of the very infrastructure of professional knowledge production – for example, intellectual property rights, open access strategies, university rankings, and indexing of journals – and the moral grounds that guide academic employment procedures and put academic mobility in perspective with regard to other forms of migration.

References

Adey, P., Bissell, D., Hannam, K., Merriman, P., & Sheller, M. (2014). Introduction. In P. Adey, D. Bissell, K. Hannam, P. Merriman, & M. Sheller (Eds.), *The Routledge handbook of mobilities* (pp. 1–20). London: Routledge.

Alexander von Humboldt Foundation – AvH (n.d.). Philipp Schwartz Initiative. Retrieved from https://www.humboldt-foundation.de/en/app ly/sponsorship-programmes/philipp-schwartz-initiative#h14796

Bakewell, O. (2008). Research beyond the categories: The importance of policy irrelevant research into forced migration. *Journal of Refugee Studies, 21*(4), 432-453. https:// doi:10.1093/jrs/fen042

Balibar, É. (2002). *Politics and the other scene.* London/New York: Verso.

Bauche, M. (2017). *Medizin und Herrschaft. Malariabekämpfung in Kamerun, Ostafrika und Ostfriesland (1890–1919).* Frankfurt/New York: Campus Verlag.

Bojadžijev, M., & Römhild, R. (2014): Was kommt nach dem „transnational turn"? Perspektiven für eine kritische Migrationsforschung. *Berliner Blätter, 65,* 10–24.

Braun, K., Georgi, F., Matthies, R., Pagano, S., & Schwertl, M. (2018). Umkämpfte Wissensproduktionen der Migration. Editorial. *movements. Journal for Critical Migration and Border Regime Studies, 4*(1), 9–27.

Brubaker, R. (2013). Categories of analysis and categories of practice: A note on the study of Muslims in European countries of immigration. *Ethnic and Racial Studies, 36*(1), 1–8. https://doi.org/10.1080/01419870.2012.729674

Carstensen, A.L., Heineshoff, L.M., Jungehülsing, J., Kirchhoff, M., & Trzeciak, M. (2014). Forschende Aktivist_innen und Aktivistische Forscher_innen: eine Hinleitung. In L.-M. Heimeshoff, S. Hess, S. Kron, H. Schwenken, & M. Trzeciak (Eds.), *Grenzregime II. Migration, Kontrolle, Wissen. Transnational Perspektiven* (pp. 257–268). Berlin/Hamburg: Assoziation A.

Casas Cortes, M., & Cobarrubias, S. (2018). It Is obvious from the map! Disobeying the production of illegality beyond borderlines. *movements. Journal for Critical Migration and Border Regime Studies, 4*(1), 29–44.

Chakrabarty, D. (2008). *Provincializing Europe. Postcolonial thought and historical difference*. Princeton: Princeton University Press.

Dahinden, J. (2016). A plea for the 'de-migranticization' of research on migration and integration. *Ethnic and Racial Studies, 39*(13), 2207–2225. https://doi.org/10.1080/01419870.2015.1124129

De Genova, N., & Peutz, N. (2010). *The Deportation regime. Sovereignty, space, and the freedom of movement*. Durham: Duke University Press.

EURAXESS – Researchers in Motion. (n.d.). Welcome refugee researchers and students looking for a suitable job! Retrieved from https://euraxess.ec.europa.eu/jobs/science4refugees

European University Association – EUA (2020). Researchers at risk: Mapping Europe's response. Report of the InspirEurope project. InspirEurope. Retrieved from https://eua.eu/downloads/publications/inspireurope%20report%20researchers%20at%20risk%20-%20mapping%20europes%20response%20final%20web.pdf

Fassin, D. (2009). Moral economies revisited. *Annales. Histoire, Sciences Sociales, 64*(6), 1237–1266.

Fassin, D. (2016). Vom Rechtsanspruch zum Gunsterweis: Zur moralischen Ökonomie der Asylvergabepraxis im heutigen Europa. *Mittelweg, 36,* 62–78.

Glick Schiller, N., & Salazar, N. B. (2013). Regimes of mobility across the globe. *Journal of Ethnic and Migration Studies, 39*(2), 183–200. https://doi.org/10.1080/1369183X.2013.723253

Hess, S., & Schmidt-Sebdner, M. (2021). Perspektiven der ethnographischen Grenzregimeforschung: Grenze als Konfliktzone. *Zeitschrift für Migrationsforschung, 1*(1), 197–214.

Horvath, K., Amelina, A., & Peters, K. (2018). Re-thinking the politics of migration. On the uses and challenges of regime perspectives for migration research. *Migration Studies*, 5(3), 301–314. https:// doi:10.1093/migration/mnx055

Hsu, M. Y. (2015). *The good immigrants. How the yellow peril became the model minority*. Princeton: Princeton University Press.

Kasparek, B., & Speer, M. (2013): At the nexus of academia and activism. Bordermonitoring.eu. *Postcolonial Studies*, 16(3), 259–268.

Krohn, C.-D. (1993). *Intellectuals in exile. Refugee scholars and the New School for Social Research*. Amherst: University of Massachusetts Press.

Lee, E. (2007). *At America's gate. Chinese immigration during the exclusion era, 1882–1943*. Chapel Hill/London: The University of North Carolina Press.

Lee, E. (2008). A nation of immigrants and a gatekeeping nation. American immigration law and policy. In R. Ueda (Ed.), *A Companion to American Immigration* (pp. 5–35). Hoboken: John Wiley & Sons.

Löhr, I. (2021): Refugee scholars as employees: Connecting the history of forced migration of scholars with the global history of higher education. In L. Dakhli, P. Laborier, & F. Wolff (Eds.), *Scholars at risk: History and politics of the protection of endangered scholars*. Wiesbaden: Springer (in print).

Löhr, I., & Reinecke, C. (2020, October 27). Not a given object: What historians can learn from the reflexive turn in migration studies. *Migrant Knowledge*. Retrieved from https://migrantknowledge.org/2020/10/27/not-a-given-o bject/

Malkki, L. (1995). Refugees and exile. From "refugee studies" to the national order of things. *Annual Review of Anthropology*, 24, 495–523.

Mezzadra, S., & Neilson, B. (2013). *Border as method, or, the multiplication of labor*. Durham: Duke University Press.

Nieswand, B. (2021). Konturen einer Moralsoziologie der Migrationsgesellschaft. *Zeitschrift für Migrationsforschung*, 1(1), 75–95.

Pott, A., Rass, C., & Wolff, F. (2018). *Was ist ein Migrationsregime? What is a migration regime?* Wiesbaden: Springer.

Randeria, S, & Römhild, R. (2013). Das postkoloniale Europa: Verflochtene Genealogien der Gegenwart – Einleitung zur erweiterten Neuauflage (2013). In S. Conrad, S. Randeria, & R. Römhild (Eds.), *Jenseits des Eurozentrismus. Postkoloniale Perspektiven in den Geschichts- und Kulturwissenschaften* (pp. 11-31). Frankfurt/New York: Campus Verlag.

Scheel, S., & Ustek Spilda, F. (2019). The politics of expertise and ignorance in the field of migration management. *Environment and Planning D: Society and Space, 37*(4), 663–681. https://doi: 10.1177/0263775819843677

Schinkel, W. (2017). *Imagined societies. A critique of immigrant integration in Western Europe.* Cambridge: Cambridge University Press.

Sheller, M. (2018). *Mobility justice. The politics of movement in the age of extremes.* London/New York: Verso.

Shils, E., & Roberts, J. (2004). The diffusion of European models outside Europe. In R. Rüegg (Ed.), *A history of the university in Europe. Vol. 3: Universities in the nineteenth and early twentieth centuries (1800–1945)* (pp. 163–231). Cambridge: Cambridge University Press.

Sivasundaram, S. (2010). Sciences and the global. On methods, questions, and theory. *Isis, 101*(1), 146–158.

Stierl, M. (2020). Do no harm? The impact of policy on migration scholarship. *Environment and Planning C: Politics and Space,* 1–20. https://doi.org/10.1177/2399654420965567

Vatansever, A. (2018). Academic nomads. Changing conceptions of academic work under precarious conditions. *Cambio, 8*(15), 153–165. https://doi.org/10.13128/cambio-22537

Walters, W. (2015). Reflections on migration and governmentality. *movements. Journal for Critical Migration and Border Regime Studies, 1*(1), 1–25. http://movements-journal.org/issues/01.grenzregime/04.walters--migration.governmentality.html

Watenpaugh, K.; Fricke, A. L., & King, J. R. (2014). *The war follows them. Syrian university students and scholars in Lebanon.* New York: Institute of International Education.

Wessels, J. (2021, January 5). The new pact on migration and asylum: Human rights challenges to border procedures. FluchtforschungsBlog. Retrieved from https://blog.fluchtforschung.net/the-new-pact-on-migration-and-asylum-human-rights-challenges-to-border-procedures/

Wonders, N. A., & Jones, L. C. (2019). Doing and undoing citizenship: The multiplication of citizenship, citizenship performances, and migration as social movement. *Theoretical Criminology, 23*(2), 136–155.

Ye, W. (2001). *Seeking modernity in China's name. Chinese students in the United States, 1900–1927.* Stanford: Stanford University Press.

Zetter, R. (2007). More labels, fewer refugees: Remaking the refugee label in an era of globalization. *Journal of Refugee Studies, 20*(2), 172–192.

Zucker, N. L., & Flink Zucker, N. (1996). *Desperate crossings: Seeking refuge in America*. Armonk: M.E. Sharpe.

Blessing or curse?
Third-party funding as a paradoxical pull factor in forced internationalization

Aslı Vatansever

Despite its chronic shortage of permanent positions and its structural inaccessibility for Bildungsausländer[1], Germany represents one of the favorite destinations for displaced scholars within the EU due to the excess of third-party funding opportunities. However, what appears as a blessing at first sight turns out to be a curse, for the plethora of external funding options actually indicate a decrease in public funding that resulted in a systematic decline of academic job security since the late 1990s and early 2000s.

In view of the rapid precarization of academic employment in Germany within roughly two decades, this chapter discusses the contradictions of forced academic migration to Germany. It argues that, with the ongoing influx of displaced scholars into its highly insecure and polarized academic labor market, Germany is becoming a reservoir of disposable academic labor in the global sense, which fails to provide viable career prospects and long-term stability to displaced academics. The analysis draws attention to the structural obstacles that hamper the development of permanent solutions to the uncertainties accompanying displacement. By doing so, the chapter aims to expand the discourse on forced academic migration from a political-economy perspective.

Following a brief presentation of the problem background, this chapter will first try to situate the issue of forced academic migration in researcher mobility studies. Germany's appeal as a host country will be explained in view of the plethora of third-party funding options. In order to elucidate

---

1   *Bildungsausländer* refers to academics/students who received their entry qualification in a country other than the country they are currently working/studying in.

the contradictory background of the excess of third-party funding schemes, an overview of the structural peculiarities of the German academic structure will be provided. Next, the risk-based temporary funding schemes designed specifically for displaced scholars will be taken under closer scrutiny. Their advantages and disadvantages will be discussed in terms of political safety and career advancement – arguably the two main concerns that impel threatened scholars to emigrate in the first place. Then, the systemic parallels and differences between the domestic academic precariat and the emigrated scholars will be discussed. The analysis will conclude by pointing out the tensions between the existing rescue programs and the structural realities of the host academic labor markets.

Problem background

As one of the leading countries in scientific production that invests 2.9% of its GDP in research and has the world's 4th strongest higher education system, Germany ranks high in terms of attracting foreign students as well as international academics and researchers (DAAD, 2020, p. 7; QS, 2018; Raupach et al., 2014, p. 7; study.eu, 2018; Wilde, 2018). At the height of the anti-intellectual attacks on academic freedoms in several countries in recent years, its appeal has extended to endangered scholars as well. Currently, Germany ranks as the top host country for threatened scholars seeking assistance from the Scholars at Risk (SAR) Network, the majority of whom come from Turkey (SAR, 2019, pp. 4–5). It is also the favorite destination within the EU, and the third favorite host country in the 2019 global ranking, for Scholar Rescue Fund (SRF) fellows from across the globe (SRF, 2020). An increasing number of threatened foreign researchers seek refuge in Germany in the hope of pursuing their research activities in a politically safe and financially stable environment. However, with 92% of its domestic academic workforce in fixed-term employment, the German academic labor market evidently suffers from a chronic shortage of permanent jobs (BMBF, 2021, p. 111).[2] As such, it is far from providing stable career advancement options even to

---

2   According to the latest federal report on early-career researchers published in 2021, the percentage of the temporarily employed academic workforce in Germany has dropped from 93% to 92% since 2017 (BMBF, 2017, pp. 29–30).

Bildungsinländer and researchers with German nationality, let alone exiled scholars with foreign nationalities and mostly non-German degrees.

This chapter argues that, in the face of the chronic lack of permanent academic positions, risk scholarships and exclusive funding schemes mainly serve to keep the displaced scholars at the margins of the host academia, without fully admitting them in. Short-term grants enable the displaced scholars to circumvent for a limited time the immediate threats such as total deprivation and deportation that await them "outside." Yet, being kept in a peculiar status between "guest" and "exile," without a permanent affiliation and access to the institutional decision-making mechanisms, they are not truly "inside" the host academic system either.

At another level, the subtle exclusion of refugee/displaced academics also seems to be in line with the neoliberal immigration policies that aim at producing conditional and fragile statuses through fixed-term employment, whereby job precarity and citizenship/residence uncertainty mutually reinforce one another (Choonara, 2020, pp. 437–438). A detailed analysis of Germany's immigration policies exceeds the scope of this chapter. However, it is important to note that in 2018 alone, the country reported the largest number of immigrants (Eurostat, 2020) and is currently trying to cope with the challenge through targeted immigration policies. The new measures include, among others, the Skilled Immigration Act of March 2020 which involves a fast-tracked procedure for qualified foreign workers from specific sectors only. Another step in the same vein was the introduction of mandatory regional dispersal mechanisms for asylum seekers according to the demographic and labor market needs of different regions (Degler & Liebig, 2017, pp. 49–50). At the same time, stricter rules for refugees entering Germany have been introduced, including temporary statuses as "tolerated persons," mandatory stays at the so-called Anchor Centers before admission, and legal consequences for not attending integration classes and skills assessment programs. On the whole, these new regulations are said to effectively replace long-term humanitarian permits with more economically oriented temporary ones (European Union Agency for Fundamental Human Rights, 2019).

Evidently, migrant labor is selectively integrated into the German labor market. Unless it features outstanding expertise in medicine, computer science, software development, or engineering – reportedly the most sought-after skills in the German labor market (DW, 2019) – migrant labor is overrepresented either in low-skill jobs or within the high-skilled yet underemployed and disposable intellectual/creative labor force. The displaced/refugee scholars clearly fit into the latter category. With an already massive oversupply of qualified academic workers Germany (or any other leading scientific host country, for that matter) obviously does not need more researchers unless they prove outstanding in their respective fields. However, an internationally outstanding track record cannot be realistically expected from the majority of displaced scholars, most of whom are coming from peripheral countries in the bottom bracket of scientific freedoms and per capita research expenditures.

There is certainly no official ban preventing exiled academics from applying for competitive federal or European grants, or even for permanent professorships at higher education institutions. Yet, as neither their previous track records achieved in their home countries nor their publications in their native language hold any significant value in the host academia, their chances are significantly lower in comparison with their German counterparts (Schmermund, 2020). A discussion on the inherent inequality of the world-systemic structures of knowledge production would go beyond the topic of this chapter. Nevertheless, it is essential to note that the labor market integration of the displaced scholars is partly hampered by the disparity between the qualification standards in the core countries and the majority of displaced academics' previous academic formation.[3]

Considering this competitive disadvantage, the relatively non-competitive risk scholarships provide a convenient preliminary entry into the host academia. However, in the face of the structural rigidity of the host academic labor markets, the exclusive risk scholarships that are supposed to represent an initial steppingstone inevitably turn into the only option for refugee/migrant scholars in the foreseeable future. For the majority, obtaining successive temporary risk scholarships becomes the only viable career option until they reach the eligibility time limit. One incidental consequence of this is that the sphere of risk scholarships turns into a sort of "quarantine section" for displaced scholars, keeping them at the outskirts of the host academic labor market. To the extent that they find themselves

---

3   Accepting the said discrepancy as default, some academic institutions even offer "post-qualification" (*Nachqualifizierung*) support to help refugee scholars catch up with their counterparts coming from a Western/European qualification background. The ProSalamander Project at the University of Duisburg-Essen is a case in point: https://www.uni-due.de/diversity/prosalamander.php.

caught in a spiral of precarious research funding with no concrete possibility of career advancement, the displaced scholars represent yet another – even more precarious – segment of the ever-growing disposable academic workforce in Germany (Vatansever, 2020a, p. 46-50).

The disadvantageous position of displaced scholars in host countries is the result of both the global inequalities over the longue durée and the decrease in academic job security in neoliberal times. Thus, the problem of integrating displaced scholars naturally exceeds the mandate of humanitarian rescue and outreach initiatives. A labor market perspective on the concrete career chances of the refugee/migrant academics is urgently needed. Moreover, the structural problems of the host academia that greatly, and almost equally, influence the occupational futures of both the native early-career staff and the exiled scholars need to be taken into account. A realistic assessment of the host environment's capacity to accommodate newcomers clearly requires an exploration of the initial pull factors as well as their implications in the mid- to long-term. For this purpose, the following section will address the shortcomings of the extant researcher mobility literature in explaining the pull factors in forced academic migration. It will attempt to provide a preliminary understanding of the possible pull factors that might paradoxically increase certain host countries' attractiveness despite their tight labor markets and exclusionary academic systems.

Forced academic migration and Germany's appeal as a host country

Due to the increased attacks on dissident scholars and scholarly communities in a number of countries, the issue of forced academic migration has recently experienced a revival of interest. However, when it comes to explaining the displaced scholars' motivations for choosing a certain host country or tracking their post-emigration career chances, two sets of problems occur. The first set concerns the discursive parameters of forced academic migration; the second set involves the limited accessibility to concrete data on refugee/migrant scholars.

The problems on the discursive level mainly stem from the analytical limitations of the notions of "academic freedom" and "scholar at risk." The discourse on forced academic migration is dominated by a humanitarian tone that insists on a human rights-oriented definition of academic freedoms (Euben, 2002; EUA, 2020, p. 9; Kinzelbach et al., 2021; Quinn, 2004). This has led to an overemphasis on the threats left behind at the country of origin, whereby the severe job and citizenship precarity that the displaced scholars face in the host countries is often ignored. Similarly, the question as to what degree academic freedom can be maintained in the absence of job security even in the host countries remains underexamined. Under these circumstances, it becomes difficult to overcome glorified portrayals of displaced scholars as "saved" intellectuals and to map out their socioeconomic coordinates as members of the academic workforce within the host labor market. The analytical and political implications of the prevalent scholars at risk discourse have been discussed thoroughly elsewhere and would exceed the scope of this chapter (Vatansever, 2020a, pp. 49–57, 145–146). Here, it should suffice to point out that the human rights-oriented category of "scholars at risk" proves analytically empty insofar that it exempts displaced scholars from academic employment relations and labor market analyses. By doing so, it also forecloses any possibility for collective action and solidarity between the native academic precariat and the migrant scholars.

The problems related to data accessibility, on the other hand, can be ascribed to the predominance of voluntary forms of academic mobility in the literature. In relation to that, there appears to be no clear distinction between voluntarily mobile international researchers and exiled academics in Germany in the official records, but only a lump categorization as "international academic staff" or "visiting researcher" (DAAD, 2019, p. 119; 2020, p. 92). Hence, neither the motivations behind the host choice nor the concrete labor market chances of the displaced scholars in the host country can be accurately identified. For this reason, the following section will hypothetically weigh possible factors for the host country choice against each other to explain the paradoxical aspects of Germany's appeal as a host country.

Migration and mobility studies in the sphere of higher education mostly focus on international student mobility (e.g., UNESCO Global Flow of Tertiary-Level Students[4]). Comparative data on researcher mobility is lacking even when it comes to voluntary mobility. As to forced academic migration, the SAR and SRF records remain the only viable source for basic quantitative information on the major countries of origin and the most commonly preferred host countries for politically persecuted scholars (SAR, 2019; SRF, 2020). With a few exceptions, forced academic mobility is also an

---

4   See report: http://uis.unesco.org/en/uis-student-flow

underrepresented topic in qualitative researcher mobility studies (Ergin et al., 2019; Laborier, 2020, p. 158; Streitwieser, 2019b). The extant qualitative literature mostly discusses the logic behind "forced internationalization" from the perspective of the higher education institutions and policymakers in the host countries (Ergin et al., 2019). According to that viewpoint, the sudden influx of refugees and migrants introduced a new motive besides the traditional academic, political, sociocultural, and economic incentives: the "humanitarian rationale" (Streitwieser, 2019a). However, since this strand represents the institutional perspective, the motivations for the displaced scholars' host country choices remain unspecified, except for a generic reference to the "mobility for survival," hinting at the threats in the home country (Özaltın et al., 2020; Streitwieser, 2019b). Yet, while the initial push factors prompting the departure may be quite obvious, the motives that lead the individual researcher or a certain group of researchers to prefer a certain host country over others, that is, the pull factors, are less so. To put it simply, the existing literature on forced researcher mobility does not provide any insight into the question of why the majority of the displaced scholars who flee to Western Europe choose to migrate to Germany instead of, for example, France or Italy.

Survey data drawn from voluntarily mobile samples identify the wish to improve one's career prospects, the need to extend professional networks, and the appeal of the host institution's reputation as the major motivations for academic mobility (DAAD, 2019, p. 115). While these and similar factors may play a certain role in the displaced academics' host choices as well, the extent to which they do so remains open to speculation. Considering the push factors in forced academic migration, it would probably not be too far off to assume that the expectations of displaced academics from a host country may first and foremost involve freedom of research, legal/political safety, and the opportunity to work. Other possible pull factors may include geographic proximity, language affinity, the inclusivity of the host academic system toward outsiders, and the existence of previous personal ties or professional networks in the country of destination (Guth, 2007; Özaltın et al., 2020, pp. 588–592; Schührer, 2018, p. 32).

With regard to the specific case of Germany as a host for displaced scholars from mostly non-European countries, however, considerations of geographic proximity are rather unlikely to weigh in against other potential hosts in close vicinity. Also, while Germany's successful record of academic freedom might represent an effective pull factor, the state of academic freedom is not much different in other host countries within the European Higher Education Area (Kinzelbach et al., 2021, pp. 9, 24; Spannagel et al., 2020, pp. 14–15). On the other hand, the language factor, or the question of accessibility of the academic system are not likely to play a favorable role at all: the German language is less widely used among the international academic community in comparison to English (Weijen, 2013) and the German academic system is notoriously impenetrable for outsiders (Afonso, 2016). Thus, logically, the availability of third-party research funding options, which the displaced scholars might initially perceive as a work opportunity, is more likely to outweigh other possible factors in attracting threatened foreign scholars.

Research funding structure and academic labor market in Germany

The key to understanding Germany's paradoxical appeal as a host country despite the exclusivity of its academic labor market lies mainly in its multivariate funding structure. Both the German academic system and its funding are marked by a complex diversity. Research is conducted by a variety of "performing sectors," including the universities, government institutions, private business enterprises, and non-profit organizations, and it is financed by various "financing sectors," including the states, the federal government, foundations, the corporate industry, and the EU. The primary financing actors for the approximately 240 universities are the states. However, the real significance of third-party funds in the German academic system becomes clear once we take a closer look at how research and development is financed.

Third-party funding refers to additional funds granted by both public and private actors, including the Federal Government, the German Research Foundation (DFG), the EU, international organizations such as the OECD or the UN, and private investors. These supplementary funds are put at the disposal of universities, departments, or affiliated researchers in various forms. They include PhD/post-doc/professorial habilitation stipends, academic awards, government-funded special programs such as the Excellence Initiative or the Higher Education Pact (Hochschulpakt), and endowed chairs funded by specific foundations (DESTATIS, 2020, p. 172). The ratio of third-party funding in research spending has increased immensely over the last two decades. Currently, 46% of the research and development expenses at German universities are being funded by third-party resources

(BMBF, 2020, p. 20). Moreover, third-party acquisition is seen as an indicator of academic excellence (Dohmen, 2016, p. 112; Wilde, 2018). Federal subsidies are granted on the basis of a university's performance record, including its capacity to acquire private third-party funding. Thus, competitive advantage in the race for private external funds also predetermines access to public Grundmittel[5] to a great extent.

This peculiarly fragmented funding structure stems from the states' chronic budgetary deficits. Despite the steady increase in student enrollments,[6] state spending in higher education has been witnessing an annual decrease by over 1% for the last 20 years (Lamprecht, 2019). However, due to the "Cooperation Ban,[7]" federal subvention could not be provided in the form of direct subsidy for teaching expenses, but only through time-bound and strictly performance-oriented allocation models such as Hochschulpakt or the Excellence Initiative. The result was an increasing reliance on external program and project funding, even for basic ongoing tasks at the universities (Meurer, 2014, p. 4). Chronically underfinanced and less career-boosting activities such as undergrad teaching and mentoring have been systematically outsourced to non-tenured staff, while the tenured staff, consisting of full professors who only make up 8% of the entire academic workforce, has been granted the luxury to focus on more lucrative and prestigious research activities that exceed basic funding. In the last two decades, the third-party revenues at universities have increased by more than 150% and, currently, almost half of the academic staff at universities is engaged in third-party projects (BMBF, 2017, p. 51; BMBF, 2021; Wilde, 2018).

The excessive third-party dependency has had adverse implications for both academic employment relations and the higher education infrastructure. The overhead costs required for the execution of third-party funded projects are often recompensed from the same, limited basic funds that higher education institutions urgently need for teaching and

---

5   *Grundmittel* refers to basic public funding provided by the federal states.

6   Plümber and Schneider (2007) provided a longitudinal analysis of the rise in student enrollment numbers. They linked the increasing enrollment rates to "fiscal opportunism" on the part of the German states. According to this, the aim of promoting student enrollments was part of a deliberate policy to decrease unemployment without increasing the budgetary resources for higher education.

7   In Germany, to preserve the federal states' autonomy in matters of research and education, cooperation between the federal and state governments is only allowed as an exception, and on a temporary basis. Cooperation Ban refers to this regulation.

administrative costs (Meurer, 2014, p. 5). Meanwhile, the factual decrease in public funding, combined with a steady annual increase by 2% in PhD production over the last two decades, has created a structural bottleneck: statistically, only about 5% of PhD holders can expect to achieve tenure (BMBF, 2017, pp. 58-60; DGB, 2020, p. 37). In the meantime, drop-out rates at the post-doc level are increasing – not the least due to the contradictory time limit on temporary employment imposed by the Fixed Term Academic Employment Law (BMBF, 2017, p. 54; Gewinner, 2018, p. 498; Raupach et al., 2014, p. 7).

Third-party funds fluctuate immensely. Research positions get canceled when projects are waived off or state funding is required to maintain permanent positions (Dilger, 2017, p. 3). The contract duration in more than half of the third-party funded temporary positions is less than a year (BMBF, 2017, pp. 29ff.; Ullrich, 2016, p. 390). Hence, the non-tenured qualification period often entails an involuntary "precarious mobility" in terms of positions, institutions, and locations (Müller & Speck, 2016; Reitz et al., 2019; Sander, 2012; Ullrich & Reitz, 2018, p. 23). During the so-called qualification period, non-tenured researchers, typically infantilized as "early-career" despite their professional experience, find themselves subordinate to the monocratic authority of full professors (Ullrich, 2016, p. 397; Vatansever, 2020b, p. 217). On the whole, the quasi-feudal hierarchies based on a deeply undemocratic notion of professorial privileges and the ever-growing project industry run by third-party funds create a precarity trap and guarantee a reservoir of disposable academic labor (Gallas, 2018; Reitz et al., 2019). In this system, where 92% lack job security and around 95% of the qualified academic workforce is statistically excluded from future permanent employment, career advancement and long-term safety for scholars with a forced migration background seem practically unachievable. Thus, exiled scholars in Germany are effectively incorporated into the reserve army of a precarious academic workforce.

Risk scholarship dependency and career instability of displaced scholars

The funding options for exiled academics represent a minor section within the gigantic third-party funding sector in Germany. Yet, they present most of the impediments of third-party overdependence, including lack of social securities, indefinite suspension of long-term planning, and forced "hyperflexibility" and "precarious mobility" (Schmid & Ullrich, 2018, p. 240; Ullrich, 2016, p. 396). In addition to the common hazards of a self-employed or third-party-funded career, the unstable nature of temporary funding poses further risks for exiled scholars – especially in terms of career advancement and fundamental securities. Juxtaposed with the push and pull factors behind forced migration, the additional risks faced in the receiving country leave room for questioning how "safe" the host environment truly is in the long run.

The main providers of risk scholarships in Germany are the Philipp Schwartz Initiative (PSI) of AvH, Baden-Württemberg Fund for persecuted scholars, the Volkswagen Foundation's Funding for Refugee Scholars and Scientists, and the Einstein Foundation's special funding program to foster academic freedom. The latter is financed by the State of Berlin and was brought to a halt until further notice due to the COVID-19-related budget cuts in 2020, only to be resumed as of January 2021, which can be viewed as an example of the fluctuating nature of third-party funding. In some cases, SAR member higher education institutions provide joint funding with the Institute of International Education Scholar Rescue Fund (IIE-SRF). Political foundations, such as the Friedrich Ebert Foundation and the Rosa Luxemburg Foundation, have also granted a limited number of irregular funds for exiled scholars over the last four years since hundreds of academics from various universities in Turkey signed the Academics for Peace Petition and faced repercussions including disciplinary investigations, dismissals, and ban from public service per decree. Some universities like Potsdam University and research institutes such as the Centre Marc Bloch and Zentrum Moderner Orient in Berlin sporadically offer individual short-term fellowships to early-career researchers or PhD students with forced migration backgrounds as well. In addition, non-profit associations provide funding for displaced/exiled scholars seeking refuge in Germany. Among them are the Off-University, which was founded by a group of persecuted and/or exiled scholars in Germany in 2017, and the Academy in Exile, which is a platform for threatened scholars maintained by a joint funding scheme of various academic institutions and foundations (EUA, 2020, p. 22).

The duration of the fellowships usually varies from three months (e.g., the Mobility Stipend offered by the Centre March Bloch in 2017) to two to three years at the most (e.g., the PSI scholarship of the AvH). Minor differences in the stipend amount and duration notwithstanding, all these scholarships have in common a rather philanthropic view on the issue of displaced academics, as is reflected in their relatively less competitive admission criteria. However, due to the lack of permanent basic funding in general, and probably also in accordance with the general immigration rules in Germany, the risk scholarships are designed in a way reminiscent of toleration periods (Duldung). Underlying this is the assumption that a return to the home country is temporarily not feasible, but that the obligation to leave persists. Theoretically, the scholarships' short temporal span suggests that they are supposed to function either as a short waiting period until the situation in the home country de-escalates or as an initial springboard into the host academia. However, since a concrete bridge to regular employment and social security factually does not exist, the risk scholarships imply a tolerating attitude toward displaced scholars, motivated by humanitarian concerns only. In other words, the host academia indicates that displaced academics are hosted only as an act of philanthropic generosity, and there exists no interest in building long-term academic collaborations with the hosted scholar.

As explicated in the previous section, the temporary nature of the risk scholarships is in line with the third-party funding industry itself and stems mainly from the structural impasse of the academic labor markets in the receiving country. The lack of future prospects in the post-scholarship phase certainly cannot be ascribed to scholarship-granting organizations or host institutions but must be considered within the general frame of employment relations in the host academia (Laborier, 2020, p. 170). However, from the perspective of the hosted scholars, the main concerns that urged them to emigrate, that is, citizenship and employment precarity, do not seem to be eliminated through short-term research funding. Unlike their native counterparts in the receiving country, who at least have access to the basic securities resulting from a stable citizenship status, exiled scholars often face multiple additional problems due to the inherent instability of a precarious career dependent on third-party funding. Contrary to the romantic portrayals of intellectual dissent and exile in the literature, those additional problems are in reality as trivial as exclusion from proper unemployment money and the inability to extend one's residence permit once the scholarship expires (Herzog & Yaka, 2019).[8] Thus, the highly unstable nature of temporary

---

8   Recently, there have been attempts to circumvent at least some of the major problems such as the refugee/migrant scholars' limited access to social rights. An important step was taken by the PSI as of early 2021 through a new program line that offers work

scholarships and the resulting hazards related to the residence status inevitably cast doubt upon the effectiveness of the contemporary rescue industry to provide a sustainable solution.

As mentioned previously, the growing academic precariat in Germany and the refugee scholars are united in the precariousness of their career prospects. However, language barriers, as well as the differences in the qualification sequences between the country of origin and the host academic environment, reduce the displaced scholars' chances even more significantly vis-à-vis the domestic surplus labor force. More importantly, refugee/migrant scholars lack the necessary social networks that are crucial in a closed and nepotistic academic system like that of Germany. The German academic system is characterized by a mixture of persistent quasi-feudal hierarchies and neoliberal hyper-competitiveness (Ullrich, 2019, pp. 156–158). Due to the unparalleled authority of the full professors and the entrenched favoritism, the career chances of junior scholars also depend – in addition to their own achievements – largely on the prestige of their mentors and their access to influential academic networks (Dorenkamp & Süß, 2017). In a cutthroat labor market highly influenced by patronage relations, the exiled scholars, who come from completely different backgrounds and lack the support of a well-connected "patron," clearly do not possess the necessary competitive edge (Gewinner, 2018, pp. 503–504; Ullrich, 2019, p. 156).

What the displaced scholars lack in terms of social capital and academic accreditation as compared with their German counterparts seems to be compensated for by the relatively non-competitive risk scholarships. For many, the SAR connection indeed represents the only access to the host academia. However, the utility of SAR-related grants is overshadowed by their long-term cost in terms of career advancement. Located in the gray zone between research funding and humanitarian help, risk scholarships are hardly ever seen as a sign of academic merit. It is "an open secret" that, in terms of academic quality and employability, a series of successive risk scholarships counts as a "negative credential" and is likely to diminish the scholarship holders' chances in the actual meritocratic labor market. In this sense, it can be argued that the risk scholarships constitute an even more insecure segment of the third-party funding industry and function like a "quarantine zone" for displaced scholars on the fringes of the academic labor markets.

---

contracts to scholarship holders and proposes to turn ongoing scholarships into work contracts.

Conclusion

This chapter departed from the assumption that Germany's popularity as a host country for displaced scholars is paradoxical, considering its closed and precarious academic labor market. In the absence of comprehensive empirical data on the pull factors for forced migration, the chapter briefly assessed possible factors that might explain Germany's attractiveness. Especially in the case of Peace Academics, who constitute the largest group of displaced academics in Germany (SAR, 2019, pp. 4–5), political safety and work opportunities are assumed to be the main motives for forced migration. Consequently, the plethora of research funding options has been viewed as the main pull factor. However, in view of the problematic correlation between the growing overreliance on third-party funding and the systematic decrease in basic funds in the last two decades, this chapter argued that the temporary research scholarships that attract an increasing number of displaced scholars to Germany will eventually prove detrimental to their long-term career prospects. Considering that, in most cases, the residence status also depends on the funding situation, the host environment is unlikely to meet exiled academics' expectations in terms of political/legal stability, job security, and career advancement in the long run.

In view of the objective difficulties of entering a highly competitive new academic environment, the initial function of risk scholarships as a sort of "boarding aid" is deemed valuable and necessary. However, as the analysis revealed, their long-term utility in terms of providing a bridge to stable employment in the host academic environment and, consequently, a secure place of residence is thwarted by a set of structural obstacles. The global discrepancy between the core and the periphery in terms of education and research certainly contributes to the competitive disadvantage of the displaced scholars in the host country. Equally challenging is the systematic precarization of the academic workforce in the host countries, of which the overdependence on temporary funding and contingent employment is but one facet.

With regard to the specific labor market situation in Germany, what Peter Ullrich has said about the domestic Mittelbau's[9] career chances also tallies with the exiled scholars' future prospects: "Many are sucked into

---

9   The non-tenured mid-level faculty is referred to as the *Mittelbau* in the public and academic discourse.

the academic career through the inflated post-doc-funding and third-party-craze, but the predominant majority does not have a perspective for future adherence" (Kaschuba, 2018). Similarly, the majority of displaced scholars flock to Germany due to the numerous research funding options, without knowing the problematic structural background of the excess of third-party funding and the massive labor precarity in German academia. Enticed by the easy initial access into the host academia through non-competitive risk scholarships, they quickly join the massive reserve army of disposable academic labor force in the host country (Vatansever, 2020a, pp. 46–57).

The immediate hazard the migrant/refugee scholars face in the receiving country is not political execution, but economic precarity and ambiguity of citizenship status. The political background of their situation notwithstanding, their current coordinates within the academic labor market conjoins them with the reserve army of disposable academic workforce in the host country, albeit with an additional degree of precarity due to the conditionality of their residence statuses. Many of them are about to hit the boundaries of the hypercompetitive and extremely precarious academic labor market in Germany and, subsequently, face all the legal obstacles tied to citizenship/residence precarity as soon as the option of risk scholarships is exhausted for various reasons. As such, displaced academics represent an exceptionally vulnerable segment within the precarious surplus labor force in the host country. Under these circumstances, the persistent "scholars at risk" label only serves to marginalize them further and perpetuate their status as "outsiders." Moreover, it conceals the parallels between the hopelessness of their occupational futures and the precarity of the domestic research staff off the tenure track.

It is clear that these structural problems exceed the mandate of outreach networks and cannot be eliminated through symptomatic solutions like short-term scholarships. Neither should the responsibility to solve them fall upon rescue programs, supporting organizations, and host institutions. Immediate interventions and quick responses to forced academic migration like risk scholarships are certainly needed. However, their effectiveness is seriously hampered by global inequalities and neoliberal employment policies in the academic sector. As a result, not only are the host countries with a high level of academic precarity like Germany turning into a repository of global surplus academic labor force, but the displaced scholars also find themselves entangled in different aspects of the same systemic problems they were intending to escape.

Under the current global-political circumstances, forced academic/intellectual migration does not seem likely to subside in the near future. The displaced scholars' stays in the host countries can be expected to extend as well. Therefore, the additional "risks" they face in the host country – such as involuntary career dropout, exclusion from social rights, and forced return/deportation – need to be addressed. A long-term solution to these problems will certainly require the academic community and policy-makers to go beyond the humanitarian conceptions of "academic freedom" and "endangered scholar," and to view the displaced scholars as a part of the academic workforce in the host country.

References

Afonso, A. (2016). Varieties of Academic Labor Markets in Europe. *PS: Political Science and Politics*, 49(4), 816–821.

Bundesministerium für Bildung und Forschung – BMBF (2017). Bundesbericht wissenschaftlicher Nachwuchs (BuWiN) 2017. Statistische Daten und Forschungsbefunde zu Promovierenden und Promovierten in Deutschland. Retrieved from https://www.buwin.de/

Bundesministerium für Bildung und Forschung – BMBF (2020). Daten und Fakten zum deutschen Forschungs- und Innovationssystem. Bundesbericht Forschung und Innovation 2020. Retrieved from https://www.bundesbericht-forschung-innovation.de/files/BMBF_BuFI-2020_Datenband.pdf

Bundesministerium für Bildung und Forschung – BMBF (2021). Bundesbericht wissenschaftlicher Nachwuchs (BuWiN) 2021. Statistische Daten und Forschungsbefunde zu Promovierenden und Promovierten in Deutschland. Retrieved from https://www.buwin.de/dateien/buwin-2021.pdf

Choonara, J. (2020). The Precarious Concept of Precarity. *Review of Radical Political Economics*, 52(3), 427–446.

Degler, E., & Liebig, T. (2017). *Finding their Way. Labor Market Integration of Refugees in Germany*. OECD International Migration Division, Directorate for Employment, Labor and Social Affairs. Retrieved from https://www.oecd.org/els/mig/Finding-their-Way-Germany.pdf

DESTATIS – Statistisches Bundesamt (2020). Bildung und Kultur. Finanzen der Hochschulen 2018. Fachserie 11, Reihe 4.5. Retrieved from https://ww

w.destatis.de/DE/Themen/Gesellschaft-Umwelt/Bildung-Forschung-Kul
tur/Bildungsfinanzen-Ausbildungsfoerderung/Publikationen/Download
s-Bildungsfinanzen/finanzen-hochschulen-2110450187004.pdf;jsessionid
=2E4C89761C06431EBC0F5D54CE0D9A56.internet8731?__blob=publicatio
nFile

Deutsche Welle – DW (2019, December 4). Merkel warns German labor
shortage could spark business exodus. Retrieved from https://www.dw.c
om/en/merkel-warns-german-labor-shortage-could-spark-business-exo
dus/a-51676100

Deutscher Akademischer Austauschdienst – DAAD (2019). *Internationalität an
deutschen Hochschulen: Erhebung von Profildaten 2018*. DAAD Studien. Bonn:
DAAD.

Deutscher Akademischer Austauschdienst – DAAD (2020). *Wissenschaft
weltoffen. Facts and Figures on the International Nature of Studies and Research
in Germany and Worldwide*. Bielefeld: wbv.

Deutscher Gewerkschaftsbund – DGB (2020). DGB-Hochschulreport: Arbeit
und Beschäftigung an Hochschulen und Forschungseinrichtungen.
Expansion und Wettbewerb im Spiegel der amtlichen Statistik. Retrieved
from https://www.dgb.de

Dilger, A. (2017). *Vor- und Nachteile verschiedener Arten von Drittmitteln.*
Diskussionspapier des Instituts für Organisationsökonomik 4/2017.
Westfälische Wilhelms-Universität Münster. Retrieved from https://www
.wiwi.uni-muenster.de/io/sites/io/files/forschen/downloads/dp-io_04_2
017.pdf

Dohmen, D. (2016). Performance-based funding of universities in Germany
– an empirical analysis. *Investigaciones de Economia de la Educación*, (11),
111–132.

Dorenkamp, I., & Süß, S. (2017). Work-life conflict among young academics:
antecedents and gender effects. *European Journal of Higher Education*, 7(4),
402–423.

Ergin, H., De Wit, H., & Leask, B. (2019). Forced Internationalization
of Higher Education: An Emerging Phenomenon. *International Higher
Education*, (97), 9–10.

Euben, D. (2002). *Academic Freedom of Individual Professors and Higher
Education Institutions: The Current Legal Landscape*. American Association of
University Professors – AAUP. Retrieved from https://www.aaup.org/site
s/default/files/files/Academic%20Freedom%20-%20Whose%20Right%20
(WEBSITE%20COPY)_6-26-02.pdf

European Union Agency for Fundamental Human Rights – FRA (2019). Tighter laws continue to hit migrants across the EU. Retrieved from https://fra.europa.eu/en/news/2019/tighter-laws-continue-hit-migrants-across-eu

European University Association – EUA (2020). Researchers at Risk. Mapping Europe's Response. InspirEurope. Retrieved from https://eua.eu/downloads/publications/inspireurope%20report%20researchers%20at%20risk%20-%20mapping%20europes%20response%20final%20web.pdf

Eurostat (2020). Migration and migrant population statistics. Statistics explained. Retrieved from https://ec.europa.eu/eurostat/statistics-explained/pdfscache/1275.pdf

Gallas, A. (2018). Precarious Academic Labor in Germany: Termed Contracts and a New Berufsverbot. *Global Labor Journal*, 9(1), 92–102.

Gewinner, I. (2018). German early career researchers in gender studies: do networks matter? *Journal of Applied Social Theory*, 1(2), 58–82.

Guth, J. (2007, February 1). Other factors influencing the mobility of scientists and their choice of destination: mobility triggers. Bundeszentrale für politische Bildung. Retrieved from https://www.bpb.de/gesellschaft/migration/kurzdossiers/58133/other-factors#footnode3-3.25

Herzog, L., & Yaka, Ö. (2019, November 29). Academic precarity is bad for everyone, but it's even worse for scholars at risk. *Times Higher Education.* Retrieved from https://www.timeshighereducation.com/blog/academic-precarity-bad-everyone-its-even-worse-scholars-risk

Kaschuba, S. (2018). Über die prekären Zustände im deutschen Wissenschaftsbetrieb. Ein Interview mit Peter Ullrich. *Soziologiemagazin*, (1), 6–11. https://doi.org/10.3224/soz.v11i1.02

Kinzelbach, K., Saliba, I., Spannagel, J., & Quinn, R. (2021). *Free Universities: Putting the Academic Freedom Index Into Action.* Full report. Global Public Policy Institute – GPPi. Retrieved from https://www.gppi.net/2020/03/26/free-universities

Laborier, P. (2020). Academic Migration and "Rescue" Programs. Between Specific and Universal Programs. In R. Roth & A. Vatansever (Eds.), *Scientific Freedom under Attack. Political Oppression, Structural Challenges, and Intellectual Resistance in Modern and Contemporary History*, (pp. 157–172). Frankfurt/New York: Campus.

Lamprecht, M. (2019, May 4). Qualitätspakt Lehre weiterentwickeln, Lehre stärker wertschätzen. FZS – freier Zusammenschluss von

StudentInnenschaften. Retrieved from https://www.fzs.de/2019/05/04/q ualitaetspakt-lehre-weiterentwickeln-lehre-staerker-wertschaetzen/

Meurer, P. (2014). Diskussion zur Weiterentwicklung des deutschen Wissenschaftssystems. Studien zum deutschen Innovationssystem No. 15. Expertenkommission Forschung und Innovation – EFI. Retrieved from https://www.e-fi.de/fileadmin/Assets/Studien/2014/StuDIS_15_201 4.pdf

Müller, A., & Speck, S. (2016). And the winner is… The male academy oder: Die ungleichen Auswirkungen universitärer Prekarität. *Sub\urban Zeitschrift für kritische Stadtforschung, 4*(2/3), 203–212.

Özaltın, D., Shakir, F., & Loizides, N. (2020). Why Do People Flee? Revisiting Forced Migration in Post-Saddam Baghdad. *Journal of International Migration and Integration, 21,* 587–610.

Plümber, T., & Schneider, C.J. (2007). Too much to die, too little to live: unemployment, higher education policies and university budgets in Germany. *Journal of European Public Policy, 14*(4), 631–653.

QS (2018). Higher Education System Strength Rankings 2018. Retrieved from https://www.topuniversities.com/system-strength-rankings/2018.

Quinn, R. (2004). Defending 'Dangerous' Minds. Reflections on the work of the Scholars at Risk Network. *Items & Issues – Social Science Research Council,* 5(1-2). https://web.archive.org/web/20100626011312/http://www.ssrc.org /workspace/images/crm/new_publication_3/%7B5cebcead-2d60-de11-bd 80-001cc477ec70%7D.pdf

Raupach, S. M. F., Lienhop, M., Karch, A., Raupach-Rosin, H., & Oltersdorf, K. M. (2014). "Exzellenz braucht Existenz". Studie zur Befristung im Wissenschaftsbereich: ein Beitrag zur Reform des Wissenschaftszeitvertragsgesetzes. Retrieved from http://www.perspekt ive-statt-befristung.de/Exzellenz_braucht_Existenz__online.pdf

Reitz, T., Janotta, L., & Cloppenburg J. (2019). Die Beschäftigungsmisere im ‚wissenschaftlichen Nachwuchs'. Ursachen und Alternativen. Zusammenstellung zentraler Daten basierend auf einer Reihe von Vorträgen 2018/19. Retrieved from https://www.mittelbau.net/wp-conte nt/upl„oads/2019/08/Überblick-Beschäftigungsmisere_NGAWiss.pdf.

Sander, N. (2012). *Das akademische Prekariat: Leben zwischen Frist und Plan.* Konstanz: UVK.

Schmermund, K. (2020, March 10). Früh über Job-Perspektiven sprechen. *Forschung & Lehre.* Retrieved from https://www.forschung-und-lehre.de/ politik/frueh-ueber-job-perspektiven-sprechen-2590/

Schmid, A., & Ullrich, P. (2018). Publish *and* Perish. Publikationszwänge, Selbstunternehmerische Wissenschaftssubjekte und Geschlecht. In E. Heitzer & S. Schultze (Eds.), *Chimära mensura? Die Human-Animal Studies zwischen Schäferhund-Science-Hoax, kritischer Geschichtswissenschaft und akademischem Trendsurfing.* (pp. 228–247). Berlin: Vergangenheitsverlag.

Scholar Rescue Fund – SRF (2020). IIE-SRF by the Numbers. Retrieved from https://www.scholarrescuefund.org/about-us/by-the-numbers/

Scholars at Risk – SAR (2019). Summary Report on Activities 2018-2019. Retrieved from https://www.scholarsatrisk.org/resources/summary-repo rt-on-activities-2018-2019/

Schührer, S. (2018). *Türkeistämmige Personen in Deutschland Erkenntnisse aus der Repräsentativuntersuchung „Ausgewählte Migrantengruppen in Deutschland 2015" (RAM).* Bundesamt für Migration und Flüchtlinge. Forschungszentrum Migration, Integration und Asyl, Working Paper 81. Retrieved from https://www.bamf.de/SharedDocs/Anlagen/EN/Forschu ng/WorkingPapers/wp81-tuerkeistaemmige-in-deutschland.pdf;jsession id=B42CB43CA960788036535D3673DED7E4.internet572?__blob=publicati onFile&v=4

Spannagel, J., Kinzelbach, K., & Saliba, I. (2020). *The Academic Freedom Index and Other New Indicators Relating to Academic Space: An Introduction.* Users Working Paper Series 26. The Varieties of Democracy Institute, University of Gothenburg, Department. of Political Science. Retrieved from https://www.v-dem.net/media/filer_public/f9/2d/f92d6e33-0682-4e 0d-84d1-5db9d6f2c977/users_working_paper_26.pdf

Streitwieser, B. (2019a, April 6). Humanism at the heart of international education. *University World News.* Retrieved from https://www.university worldnews.com/post.php?story=20190405132624741

Streitwieser, B. (2019b). International Education for Enlightenment, for Opportunity and for Survival: Where Students, Migrants and Refugees Diverge. *Journal of Comparative and International Higher Education, 11,* 4-9.

Study.eu (2018). The Study. EU Country Ranking 2018 for International Students. Retrieved from www.study.eu/article/the-study-eu-country-ranking-2018-for-international-students

Ullrich, P. (2016). Prekäre Wissensarbeit im akademischen Kapitalismus. Strukturen, Subjektivitäten und Organisierungsansätze in Mittelbau und Fachgesellschaften. *Soziologie, 45*(4), 388–411. http://doi.org/10.14279/depo sitonce-5919

Ullrich, P. (2019). In Itself But Not Yet For Itself – Organizing The New Academic Precariat. In W. Baier, E. Canepa & H. Golemis (Eds.), *The radical left in Europe: rediscovering hope.* (pp. 155-168). London: The Merlin Press.

Ullrich, P., & Reitz, T. (2018). Raus aus der prekären Mobilität. *Forum Wissenschaft,* 23–24.

Vatansever, A. (2020a). *At the Margins of Academia. Exile, Precariousness, and Subjectivity.* Boston/Leiden: Brill.

Vatansever, A. (2020b). Between Excellence and Precariousness: The Transformation of Academic Labor Relations in Germany. In R. Roth & A. Vatansever (Eds.), *Scientific Freedom under Attack. Political Oppression, Structural Challenges, and Intellectual Resistance in Modern and Contemporary History.* (pp. 215-226). Frankfurt/New York: Campus.

Weijen, D. (2013). Publication languages in the Arts and Humanities. *Research Trends,* 32. https://www.researchtrends.com/issue-32-march-2013/public ation-languages-in-the-arts-humanities-2/

Wilde, A. (2018). Research is key priority for politics and industry in Germany. *Die Zeit. Forschung & Lehre.* https://www.academics.com/guide/research-i n-germany.

Scholars in exile in the Netherlands
When humanitarian support encounters neoliberal reform practices

Lizzy Anjel-van Dijk & Maggi W.H. Leung

Academics in many parts of the world are facing increasingly uncertain and insecure conditions. They are threatened for their ideas, research, and social position because of the critical questions they ask. The protection of these people, many of whom are in exile, and of academic freedom in general, is of paramount importance. The right of every human being to think, ask questions, and exchange ideas is the basis for science and a free and thriving society. Hosting at-risk and exiled scholars is not only a sign of solidarity; their presence also enriches academia. This chapter discusses the Dutch academic system and charts the support programs in place in the Netherlands. We show how these programs enable highly qualified, forcibly displaced, and at-risk scholars to continue their academic careers and research activities. We do so by recounting some of their personal experiences.

In response to violations of academic freedom and to increase the diversity in Dutch academia, a number of initiatives have been established by various institutions in the Netherlands to provide support. In this chapter, we argue that these initiatives should be understood within the context of the broader, transforming academia in the Netherlands, rather than as isolated solidarity or even charity efforts. Specifically, we thematize the intensive neoliberalization of the Dutch and global academic system, which has led to a fiercely competitive environment and immense pressure to publish and compete for research grants.[1] This trend puts pressure on all scholars in

---

[1] Politicization of the Dutch academia is also an important topic, but it is beyond the scope of this chapter.

the Dutch academic field. It is, however, important to ask how it affects at-risk scholars who are already placed in disadvantaged positions in the Netherlands.

In the following, we map out the support landscape relevant to scholars at risk. While presenting the opportunities – for both the scholars and the host academia/society – we also discuss the challenges. In doing so, we rely on qualitative methods of data collection and analysis. We use data from existing reports and studies on the recent transformation of Dutch higher education and from nine in-depth semi-structured interviews that we conducted. We spoke to five at-risk and displaced scholars, as well as two people from supporting institutions (non-governmental organizations [NGOs]) and two from different universities in the Netherlands. In all nine interviews, particular attention was paid to the privacy and anonymity of the respondents and to digital security.

The interviewed scholars' backgrounds ranged from being a librarian and a women's rights activist, a professor of cultural anthropology, and an architect from the Middle East/western Asian region to a professor and researcher in the field of public international law from southern Africa. They had been in the Netherlands from two years (in the case of a scholar who felt forced to return home due to a lack of opportunities in Dutch academia) to seven years. We combined the different data sets to analyze the institutional landscape in the Netherlands and the lived experiences of exiled scholars.

The chapter is structured as follows. First, we present an overview of existing support programs in the Netherlands that host and integrate exiled and at-risk scholars. Next, we introduce the Dutch context by providing background information on the country's higher education institution (HEI) sector and its transformation over recent decades. This is followed by a discussion on the opportunities and challenges faced by at-risk scholars during their stay in the Netherlands. In doing so, we underline the challenges posed by the market-oriented Dutch higher education system to academics in general and to at-risk scholars in particular.

Support programs in the Netherlands

In response to worsening violations of academic freedom worldwide, several initiatives have been introduced by universities and NGOs in the Netherlands to support threatened and at-risk academics. These initiatives are based on the principle of international humanitarian responsibility and academic solidarity. Figure 1 illustrates some of the key support programs and institutions.

Figure 1. Relation among programs/initiatives supporting at-risk scholars in the Netherlands.

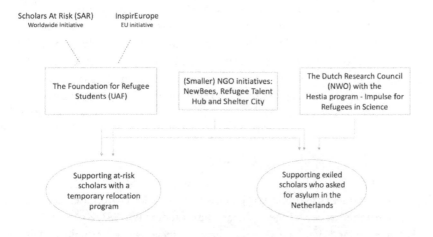

Scholars at Risk program

The Scholars at Risk (SAR) program is not a Dutch program but an international network of 440 HEIs from 40 countries. This network partners with the European University Association and has national sections in 13 European countries of which the Netherlands is one – together with Belgium, Finland, Germany, Ireland, Italy, Norway, Slovakia, Sweden, Switzerland and United Kingdom (SAR, n.d.-a). SAR is committed to protecting and supporting at-risk scholars by providing them with sanctuary and assistance. Temporary research and teaching positions, ranging from three months to two years, are arranged for threatened scholars. During this time, scholars may teach, research, lecture, or study. More than 300 researchers around the world receive support from SAR each year. In addition to helping to arrange work positions, SAR provides advisory services for scholars and their hosts. It also has an advocacy function, namely, defending threatened academics and raising awareness of their causes.

The SAR program in the Netherlands was established in 2009 by HEIs in the Netherlands and Belgium, with the purpose of offering threatened scholars (who have PhDs and teaching and/or research experience at an HEI) the opportunity to temporarily continue their work or studies in a safe environment. When the situation in the country of origin becomes or is considered safe again, the scholars are expected to return. The SAR program arranges temporary research and teaching positions at various institutions and provides advice and referral services (SAR, n.d-b; UAF, 2020).

Scholars who join the SAR programs receive at least one year of secured financial and employment conditions, including help integrating into the labor market. Our interviewees said that they appreciated the program allowing them to return to a life of relative normality after their traumatizing experiences and offering them social stability and international work experience (personal interview with at-risk scholar, December 2020; SUCRE, 2018).

Foundation for Refugee Students UAF

As an intermediary support organization, the UAF coordinates the SAR scheme and provides refugee scholars with funding (via SAR New York) and support. In addition, the UAF has been supporting, as its main activity, refugee and asylum-seeking students and professionals (including scholars) in their studies and job searches in the Dutch labor market since 1948. Support is offered through personal guidance and workshops, and the UAF provides financial assistance to and lobbies on behalf of this group. Its main activities are focused solely on refugees and asylum seekers. This means that the support receivers must have applied for asylum in Europe and obtained a Dutch residence permit.

In 2020, the UAF assisted 3,778 refugees through a modular guidance program that matched their qualifications and educational or career plans. Of this group, 47% were studying, 9% were seeking employment, and the remaining 44% were preparing for studies by doing a language course, receiving mentoring or doing a pre-bachelor program (Schakeljaar, literally "switch year"). The UAF facilitated 37 placements of at-risk scholars via the SAR program in 2020 (UAF, 2021).

Each year, the UAF helps with the placement and training of 25 to 30 people within the SAR program. Most exiled and at-risk scholars in the

Netherlands come from Syria, Iraq, Iran, or Turkey (E. Amadi Salumu, personal communication, September 4, 2020).

In 2020, the UAF and nine partner organizations across Europe launched the InspirEurope program – an initiative to support researchers who are at risk due to discrimination, persecution, or violence and to lay the foundations for a sustainable network that would increase opportunities for these researchers in academic and non-academic sectors throughout Europe. As a Europe-wide network, the program complements national programs and offers opportunities to gain new insights. For instance, it highlights the challenges faced by at-risk researchers posed by European policymaking and offers an appropriate context for solutions to address them. It also facilitates knowledge transfer between existing and new support initiatives and between academic and non-academic sectors in member countries. As a network, it offers more career prospects for at-risk researchers in the form of improved chances of receiving funding or finding employment. It also provides space for the greater involvement of central, eastern and southern European stakeholders in support measures.

The Hestia program of the Dutch Research Council (NWO)

During an interview with E. Amadi Salumu of the UAF on September 4, 2020, he explained that the overall focus in the Netherlands shifted to hosting scholars who held asylum seekers' residence permits. One of the programs targeting this group of at-risk scholars is the Hestia – Impulse for Refugees in Science program of the Nederlandse Organisatie voor Wetenschappelijk Onderzoek (NWO; Dutch Research Council). The Hestia program makes it possible for the leader and/or main applicant of a research project based at a Dutch university to apply for funding for a refugee with an academic background to join the project. Such appointments are for 18 months, or 24 months in the case of part-time appointments. The program finished its third pilot phase in 2020 and is regarded by the Research Council as an immense success:

> There is a great need to continue the program in its unchanged form because it is functioning very positively and the acquisition of a scholarship in competition has received a very positive response. After all, that is the measure of success within Anglo-Saxon academia at the moment. So this immediately shows that the laureate has the skills that are required in

the academy and that the person is competitive et cetera. Universities are therefore also very positively engaged with the laureates of the Hestia program and in many cases also offer an extension. For example, to write follow-up research applications for PhD positions. (M. van Dijk, personal communication, November 11, 2020)

During the three pilot phases, a total of 116 refugee scholars applied to join the Hestia program. Of these, 32 scholars were selected, and they and their research teams received research grants. During the same interview, we were told that almost all of the refugee scholars who were selected were able to continue their professional careers through more funding possibilities or PhD positions.

Non-governmental initiatives: NewBees, Refugee Talent Hub, and Shelter City

There are several other initiatives in the Netherlands from which at-risk or exiled scholars can benefit, even though they do not have their focus specifically on exiled scholars but more broadly on refugees. NewBees, for example, matches newcomers, including refugees, to traineeships with local entrepreneurs and organizations. Here, newcomers gain experience in workshops and practical experience in a workplace under the supervision of NewBees' matchers. Similarly, the Refugee Talent Hub organizes small-scale meetings for refugees and employers, with the aim of fostering employment. Even though these programs have proved to be valuable to their target groups, only people who have been granted asylum can participate, and the programs do not have the specific focus on or expertise in at-risk scholars aspiring to enter Dutch academia.

Another example is the Shelter City program, which offers a three-month refuge to human rights defenders who report and stand up against human rights violations in their home countries. During the program, the human rights defenders receive capacity-building tools and training and can increase their networks while advocating for their causes. The program serves human rights defenders broadly and has also supported at-risk scholars who report on human rights violations. The three-month relocation program does not intend to provide a longer-term exile but appeals to those who want to catch their breath and make connections abroad before returning home to continue their human rights work. During their stay in the Netherlands, the human rights defenders and scholars step onto the public stage: they talk to journalists and politicians and visit partner organizations and other stakeholders. Higher international visibility through these activities is intended to protect them from being threatened, intimidated, or even killed by opponents. Since the establishment of the first Shelter City in 2012, the initiative has been expanded to a network of 17 cities and has since hosted more than 200 human rights defenders.

Having sketched the support landscape in the Netherlands, we will now discuss a few of the key aspects that drew our attention while speaking to our interviewees.

Dutch academic system as the host environment

Being displaced, at-risk scholars seek a safe environment for themselves and their families, as well as an appropriate position that will allow them to continue their academic work. While driven by solidarity motivations when supporting these scholars, HEIs also gain new capacities and knowledge through the hosting arrangement. Hosting or employing highly qualified at-risk scholars can be very beneficial, leading to a more diverse and richer set of skills in specialized research teams and in HEIs in general. Yet, such a win-win situation is not always a given reality. The logic of support programs often creates an image of exiled scholars as victims who need to be saved rather than producers of knowledge contributing to the host institutions' research agendas. However, others hold a different view, as stated in the SUCRE report (2018): "HEIs are not charity organizations and threatened researchers are not simply looking for mercy." Hence, the relationships of the parties involved are more complex than it might seem at first sight, as a number of challenges may arise during the hosting cycle.

It is important to situate these relationships in the changing Dutch and global higher education context. In the Netherlands, the higher education system has two segments, consisting of university education and higher professional education institutions. There are currently 13 traditional research universities in the Netherlands that together enroll around 200,000 students annually (Maassen et al., 2011). The higher professional education sector consists of 42 institutions (hogescholen), which enroll more than 350,000 students annually (Maassen et al., 2011).

The Dutch government used to implement detailed planning and exert tight control over the inputs of institutions of higher education and their resources, students and staff. This changed in 1985 following the introduction of a new government strategy regarding higher education in the Netherlands (Hoger Onderwijs: Autonomie en Kwaliteit [Higher Education Autonomy and Quality]) (Maassen et al., 2011; van Vught, 1997). The new strategy adopted by the Ministry of Education and Science departed from the traditional centralized control model. The main aim of the reform was to strengthen the autonomy of HEIs, increase their adaptive capacity and flexibility to respond to the needs of society, and raise the levels of quality and differentiation of the system (Ministry of Education and Science, 1985). Strengthening institutional autonomy was seen as a major element in improving the functioning of the higher education system. Thus, universities also had to develop more direct links with external funding sources. HEIs are expected to better address the needs of society and – very importantly – diversify income sources, compensating for the cuts in the governmental budget for higher education.

The Dutch government's decreased role in providing higher education as a public good allowed market principles to find their way into the management of higher education. The result was an increasingly commodified higher education and research system in the Netherlands. For the last 20 years, the student population in the Netherlands has been growing, while the budgets of universities have not (ReThink UvA, 2015). It has therefore become more important to attract funding from external sources. De Boer et al. (2007) reported that from 1992 to 2003, the total revenues from activities with third parties (contract activities) increased from 548 to 1,257 million euros.

Research output, productivity, and assessment of HEIs are now being measured primarily in quantitative terms. Even though changes have been proposed and practiced in some universities to give due recognition to teaching and other forms of impact, academics are still being de facto assessed by the number of grants, amount of external funding acquired, number of (high-ranking journal) publications, citation indices, and impact factors. The Science in Transition initiative (2013), an organization consisting of Dutch scholars, has lamented that researchers have become economically dependent on their publications and that academic success depends on the number of publications instead of the societal relevance of research.

As a result of the reform, Dutch universities have been in a state of transition, with the aim of becoming the world's top "knowledge economy" (Maassen et al., 2011). This transition is characterized by immense pressure to become more efficient and produce higher outputs. Ties have been built with society and the business sector to transform the economy and to become an incubator for top academic talents. In that sense, universities have been denationalized and transformed from state agencies into public corporations (Maassen, 2008).

Although the marketplace approach has brought universities more resources, a higher number of and arguably "better" students, a larger capacity for advancing knowledge, and a more productive role in the Dutch economy, it has in many ways also diminished the sovereignty of universities over their own activities, as they have been weakened in their mission to serve the public. Universities' independence from the government has growing commercial entanglements. The new "modernized" model emphasizes leadership, management, and entrepreneurship. Research and higher education are identified as key instruments for economic performance and growth and for mastering global competition. Traditional values such as individual academic freedom and internal democracy have, on the other hand, been given less importance.

How do at-risk and displaced scholars, who are mostly placed in a less advantageous position in the system, experience this highly competitive setting? The reflections of one of our at-risk and exiled scholars illustrates this tension clearly. Our interviewee, a scholar from southern Africa,[2] shared with us his opinions on the opportunities, challenges, and contradictions in the Dutch academic system:

> Is the academic freedom here in the Netherlands good enough to host academics from other countries? Problems also exist in the Netherlands relating to the capitalization of academia but also the politicization. What is politically correct? What can you say and what can you not say? ... This affects Dutch scholars but even more us, immigrant scholars, because we are already in a disadvantaged position. (At-risk scholar in the field of cultural anthropology who has worked in Dutch academia for five years)

---

2   Due to the relatively small number of scholars in exile in the Netherlands, we have chosen not to disclose the specific backgrounds of our interviewees to ensure anonymity.

Recognition of knowledge and credentials

The criteria of academic success vary from country to country. The academic system that the exiled scholars were used to in their home countries can be different in many ways from that in the Netherlands. All the articles they wrote and all the courses they developed and taught in their home countries can become invisible in Dutch academia. To start anew, they need to compete in a mostly English-speaking environment, where the value of a researcher depends solely on a list of EU-credited degrees and certificates and the number of research papers (preferably written in English or Dutch) they publish.

This loss of earlier achievements is often overlooked by host universities, funding institutions, and foundations. Such a competitive system takes a high toll on all academics, but a much greater one on at-risk and exiled scholars who have been forced to flee political persecution in their home countries. The latter must satisfy the host country's requirements for credentials within a relatively brief period of time before they can obtain a position in Dutch academia, where their valuable (alternative) views and international experiences and skills can be recognized.

Regarding the recognition of expertise, we learned about prejudices in Dutch academia faced by our scholar interviewees. Some expressed frustration that they felt limited to only contributing to research areas related to their ethnicity:

> [w]hen there was something about Africa [at the Dutch HEI], they expected you to concentrate on this knowledge and it limits you on other possibilities that you would like to work on . . . I work in the area of public international law . . . so I want to continue developing my work in international law, and not be restricted to some elections somewhere in Africa or any situation on climate change in Kenya. I should be allowed to continue my specialization and be allowed to even challenge my international law colleagues. (At-risk scholar in the field of international public law who has worked in Dutch academia for five years).

In this way, the development of knowledge is restricted. At-risk scholars should be allowed to choose the areas they would like to work in and not be limited to a set of topics that are not part of their specialization.

Another important topic that was mentioned in the interviews was the problem of the "politicization of knowledge":

What is politically correct? What can you say and what can you not say? So that is what I see, for example, at conferences with people who are African. They are not so free to share their opinions in the plenary session. Often during break times, we have our own session where we discuss issues frankly. We then discuss how things might be wrong in the plenary session or how others might have missed the point there. During plenary sessions, it cannot feel safe enough to speak out. You feel forced to say what is politically correct and you have to be careful about what you are saying. (At-risk scholar in the field of international public law who has worked in Dutch academia for five years).

Temporal dimensions

During the personal interviews, we learned that time-related challenges are paramount among exiled scholars. These have to do with how support programs are often limited to the short term. Many of our interviewees said that when they arrived in the Netherlands, they felt pressured to immediately make plans for their future, at the time when they were both struggling with the new context and suffering from the past. There are special and very helpful scholarships and funding opportunities for academics in this situation, but they typically cover only one or two years, as is the case in many European host countries.

Most scholars have to adapt to a very different normality compared with that in their home contexts. They face intercultural challenges and unfamiliarity with local customs and the languages used in the Dutch academic system. All this is time-consuming and may overshadow the relief of having reached a safe place and their enthusiasm for working in academia again. In one interview, an exiled and at-risk scholar said that it had taken her around three years to be able to work more productively in the Netherlands:

It was a pity for me. I was in a place full of knowledge, full of academic relationships, but I was not able to make good use of this. It would be my dream to go back to [the host HEI] and make better use of this. I was going through the darkest times in my life and was even suicidal. At that time, it was not possible to take exams. (At-risk scholar and women's rights activist who has worked in Dutch academia for seven years)

Exiled and at-risk scholars need additional support to integrate into the Netherlands both during and especially after their fellowships or temporarily funded positions. They have to successfully integrate into a foreign country and academic environment, become part of new disciplinary and interdisciplinary networks, and learn to write funding applications while also publishing articles. This needs to happen in a fiercely competitive academic environment, as discussed earlier. Additionally, a contact person working with exiled scholars who joined the SAR program said during a personal interview that the main problem is often not the academic system itself, but the trauma of loss and having to start over:

> When they come to the Netherlands, they usually don't immediately start publishing, or hand in assignments or book chapters. A man who recently came to the Netherlands and who I support, only shows up for one in every three appointments since he has a lot on his mind and is still living in another world. He was a somebody in his country [lawyer] and in the Netherlands he is back at the bottom of the pile. (personal interview with Head of executive services HEI, November 4, 2020)

Therefore, it is often unrealistic, or even inconsiderate, to expect exiled scholars to be able to compete in a different academic labor market and the Dutch knowledge economy, which is highly competitive even for Dutch academics and especially for those with prior traumatizing experiences. The time limits set on them add another layer of pressure.

Inclusion beyond work

HEIs that are committed to supporting exiled and at-risk scholars usually provide services to help them enter the academic system. Some services extend beyond the work realm to enhance newcomers' settling in process. Most of these services, however, focus only on the early phase (i.e., the first 4–6 weeks), namely by providing help with visa and residency issues, housing, and registration formalities. As international offices or Human Resources departments are usually busy with many more clients, support often ends after the initial phase, and it is up to the exiled scholars to ask for additional help. This, however, requires persistence and assertiveness, which does not come naturally to everybody, especially scholars from other cultural backgrounds who are recovering from traumas.

On arrival in the Netherlands, the main concern of at-risk scholars is often housing and settling into the country and Dutch academia. Yet, the difficulty of going back to "normality" for these persecuted scholars is often underestimated. When at-risk scholars come to the Netherlands, they may have lots of expectations, but in reality, they can do very little at the beginning because they might be dealing with trauma, grief, and loss during their involuntary exile. They might experience loneliness, the feeling of being lost, and the loss of status and career prospects. Therefore, many scholars need additional support to get back on their feet. However, returning to a safe everyday life routine in the Netherlands is not only linked to bureaucratic paperwork, but can also be a huge challenge after a period of continuous unrest and traumatic experiences. Finding a skilled mentor can then be a helpful relief for exiled scholars.

At some HEIs, at-risk scholars are also introduced to a contact person who provides close supervision, answers any questions they might have, and introduces them to the Netherlands. Our interviews revealed that at-risk scholars have very different experiences in this regard. Some felt that their contact person was very hospitable and connected them to their HEI and Dutch society, even calling them their "Dutch best friend," while others had very little contact with their contact person, leaving them with a feeling of being alone and frustrated:

> Starting in the Dutch academic system was not easy. I was just given a desk and a computer and they said you can work from here ... overall I mostly worked on publications and they just left me to do what I wanted to do, so that easily leaves you frustrated ... You have to create opportunities for yourself but it would be great if you came and could immediately become part of something. (At-risk scholar in the field of cultural anthropology who has worked in Dutch academia for five years)

Special staff with sufficient time resources to follow up placements and host exiled and at-risk scholars must be appointed, and HEIs need continuous support when hosting such scholars. This underlines the importance of supporting not only these scholars but also the academic hosts who, often for the first time in their careers, are encountering the challenge of hosting people with very different sets of experiences and, most likely, who have also experienced traumas. As mentioned, all new scholars entering an HEI need some time to settle into their new environment. This group of exiled and at-risk scholars is affected by many additional challenges, such as adjusting to a new language and academic culture, dealing with their past traumas, and bringing their families over to join them. Thus, academic hosts must be made aware of the challenges and responsibilities associated with taking up the role of mentor and helping exiled scholars (and their family members) regain a sense of peace and normalcy after their traumatic experiences and to settle into a safe place and resume working in positions that match their academic training, until either a return to the home country is possible or the scholars and their families have managed to feel at home in the host country. It was mentioned in the interviews that the UAF was very good at offering support as they provided the resources to help the scholars settle into their new environment. This support is, however, limited to more practical matters, and the main task of finding a job and getting into academia is something that needs to be done at the level of the host HEI. The contact person of the HEI is therefore an essential mediator.

Financial issues

While the interviewed at-risk scholars said that they fully appreciated the privileged situation of being temporarily in a safe place and able to continue working in academia, many nevertheless suffer from their precarious status and unstable financial situation. It was mentioned that there is a feeling among migrant scholars of being underpaid:

> When I came here I was offered a position, but the money I was earning was even less than the money that I received as a stipend from the UAF. The stipend is already barely enough to survive on here, and anything less than that is just slavery. So this is something I had to discuss with my immediate supervisor and at the moment they are looking into it. . . . I don't want to say that my Dutch colleagues think that Dutch education is superior but . . . let's say the systems are different such as making use of [a web-based learning management system]. These are things that I still had to learn. So maybe they are then thinking that they have to start working with me at an entry-level position. (At-risk scholar in the field of international public law who has worked in Dutch academia for five years)

As mentioned, existing programs and funds supporting at-risk scholars can be highly lauded for their goodwill. However, academic systems such as those in the Netherlands have increasingly precarious employment situations with, for example, insufficient funding possibilities, a lack of security and short-termism. This also makes it increasingly difficult to support exiled and at-risk scholars, whose disadvantaged position often prevents them from claiming their rights:

> With people like us, we have been fighting our whole lives. And we just get tired ... as was also the case with me. When I got the contract, I just said "Okay, that is fine." I will get whatever they give me, and after I get some more experience, I will see if I can move to the next step in life. Because you also feel as if they are doing you a favor by including you in academia. So being in situations like this can also prohibit you in claiming your rights. (At-risk scholar in the field of international public law who has worked in Dutch academia for five years)

Uncertainty around support programs at risk

In April 2021, the UAF discontinued the SAR program, meaning that it will no longer support scholars who have not sought asylum in the Netherlands. A representative of a Dutch HEI (personal communication, November 4, 2020) told us that the decision was made due to the ways PhD candidates are employed by universities in the Netherlands. Instead of receiving allowances, candidates are employed by their universities at which they carry out their research and teaching duties. In general, this system offers protection for PhD researchers who are employed by universities. However, it is also a source of confusion regarding the tax situation within the framework of the SAR program. The allowances that at-risk scholars receive from the SAR program are subject to the more regularly used bursary conditions. Even though at-risk scholars who joined the SAR program have a formal guest researcher status at their HEI, the allowances they receive are not taxed, which means they are not eligible to benefit from Dutch social security schemes. Moreover, the Dutch tax authorities are often not clear about the regulations that apply to the SAR allowances. This became a burden for the universities. To relieve this burden, the UAF was chosen by SAR as a partner, so the former could be the central point in making an agreement with the tax authorities as well as providing scholars with counseling. However, in 2020, the UAF decided it was too risky to continue this structure, as it might be charged with employing at-risk scholars on a temporary basis.

There were discussions about whether universities should assume the role of the UAF, meaning that they would also need to offer guidance, provide training, and be responsible for administering the allowances. However, past experiences of similar constructs have not been positive, as this puts additional workload and pressure on the HEIs. They have to deal with the Dutch tax authority and related counseling for scholars on their own, rather than having the UAF as the central point. As this was considered to be undesirable, a long search started for another organization to implement the SAR program. In a later conversation with a colleague at the UAF (E. Amadi Salumu, personal communication, September 10, 2021), it was mentioned that Nuffic, the Dutch organization for internationalization in education, is consulting with SAR New York on its implementation in the Netherlands. The future remains uncertain.

Conclusion

Although the vision and process of internationalization in the Netherlands, along with the country's position as one of the world's top five knowledge economies (UNDP, 2020), are strongly hyped by the Dutch government, the country's academic system is far from inclusive. When confronted with the reality many exiled scholars face in the Netherlands, we might then ask whether the free academic space in the Netherlands is shrinking. As we have shown in this chapter, constraints are felt not only by academics from faraway countries: the increasingly marketized academic world, pressure to publish in high-impact journals, and competition for research grants are all harsh realities that create precarity among academics, especially young scholars in disciplines that are less well funded.

It seems that in the Netherlands, more support programs for exiled and at-risk scholars that focus on integration and offer ways for such scholars to obtain research grants have been created. However, these programs focus on at-risk scholars who hold asylum seekers' residence permits. Especially with the current problems trying to keep the SAR program active in the Netherlands, scholars who seek temporary relocation and do not want to apply for asylum, or who need to temporarily relocate to live in safety and have some time to breathe, are left behind.

There are also challenges related to finding and retaining adequate positions in the highly competitive research labor market. The limited amount of time that at-risk scholars have to carry out research and/or teach, settle into a new country professionally and privately, recover from stress and possible traumas, learn Dutch, and find follow-up employment is especially challenging. Moreover, for many exiled scholars coping with trauma is among the main challenges. Therefore, funding programs should be supplemented with mentoring programs as well as inclusive and supportive research environments. At-risk scholars need to be introduced to academic networks and receive support applying for research grants and permanent positions.

In order to enable inclusion of at-risk scholars, the structural problems of the academic system both in the Netherlands and abroad must be dealt with. Market-oriented academic systems force academics to be entrepreneurial and compete with each other, which often reduces their readiness to collaborate with fellow academics who are seen as less competitive. One of the groups of scholars who are most seriously affected by this competition and precarity in academic labor are those who are already threatened, namely at-risk and displaced scholars. This leads to two pertinent questions: How should internationalization, inclusion and solidarity be practiced? And how can we envision more genuine and inclusive knowledge co-creation? These questions are important for the sake not only of at-risk scholars but also of academia as a whole.

References

De Boer, H.F., Enders, J., & Leisyte, L. (2007). Public sector reform in Dutch higher education: The Organizational Transformation of the University. *Public Administration*, 85(1), 27–46.

Maassen P. (2008). The Modernisation Of European Higher Education. In A. Amaral, I. Bleiklie & C. Musselin (Eds.), *From Governance to Identity. Higher Education Dynamics*. Dordrecht: Springer. https://doi.org/10.1007/978-1-4 020-8994-7_8

Maassen, P., Moen, E., & Stensaker, B. (2011). Reforming higher education in the Netherlands and Norway: the role of the state and national modes of governance. *Policy Studies*, 32(5), 479–495.

Ministry of Education and Science (1985) *Hoger onderwijs: autonomie en kwaliteit [Higher Education: Autonomy and Quality]*. Zoetermeer.

ReThink UvA (2015, September). Position papers on teaching and research. Retrieved from http://rethinkuva.org/blog/2015/09/17/position-papers-on-teaching-and-research-2/

Scholars at Risk – SAR (n.d.-a). SAR sections. Retrieved from https://www.scholarsatrisk.org/sar-sections/

Scholars at Risk – SAR (n.d.-b). Get Help. Retrieved from https://www.scholarsatrisk.org/get-help/

Science in Transition (2013). Over Science in Transition. Retrieved from http://www.scienceintransition.nl/over-science-in-transition

Supporting University Community Pathways for Refugees-Migrants – SUCRE (2018). Institutional support for refugee scholars in higher education. Retrieved from https://ec.europa.eu/programmes/erasmus-plus/project-result-content/c555d3e6-6f58-4630-a593-a1d1e42cf573/IO3_Publication.pdf

UAF (2020). InSPIREurope: Initiatief ter ondersteuning, bevordering en integratie van wetenschappers. Retrieved from https://www.uaf.nl/kennisbank/inspireurope

UAF (2021). Jaarverslag 2020. Retrieved from https://www.uaf.nl/wp-content/uploads/2021/06/UAF_Jaarverslag-en-jaarrekening-2020.pdf

United Nations Development Programme – UNDP (2020). Global Knowledge Index 2020. Retrieved from https://www.knowledge4all.com/Reports/globalknowledgeindx2020_en.pdf

Van Vught, F.A. (1997). Combining planning and the market: an analysis of the Government strategy towards higher education in the Netherlands. *Higher Education Policy,* 10(3/4), 211–224.

The emergence of a "third space"?

Critical scholars from Turkey[1]
Challenges and opportunities in Germany

Ergün Özgür

In the last two decades, democracy, human rights, and freedoms, especially academic freedom, have been challenged by populist right-wing and authoritarian regimes in Turkey, and many other countries around the world. Many who have raised their voices in opposition, including academics, artists, and journalists, have been forced to leave their country due to increasing pressures, conflicts, and wars.

Since January 11, 2016, the lives of the Academics for Peace (AfP, or BAK in Turkish) from Turkey who were the signatories of the "Peace Petition" entitled "We will not be a party to this crime" have changed drastically. These scholars will be called critical scholars in this article. What characterizes them is that they have criticized the state violence in the Kurdish-populated eastern and southeastern regions of Turkey. They are critical of the assignment of rectors to the universities or trustees to the municipalities instead of elected representatives. They are against the state's pressure on elected members of the parliament and oppositional voices such as those of academics or journalists. They fight against increasing nationalism and authoritarianism and criticize the competitive liberal higher education system, which generates precarity, especially for junior scholars. Many of them are social scientists studying topics such as the Kurdish conflict, the Armenian genocide, LGBTQI+, gender, minority and human rights, migration, and refugee issues. After the public announcement of the Peace Petition in Turkey, the signatories, these critical scholars, were attacked by high-ranking state

---

1   Parts of this study were conducted during my affiliation with Leibniz-Zentrum Moderner Orient (ZMO), Berlin, and the article was written at the Institute for Media and Communication Studies of the Freie Universität Berlin as an Einstein Guest Researcher.

officials and organizations, including Turkish President Erdoğan, Prime Minister Davutoğlu, pro-government media, officials of the Council of Higher Education (CoHE: Yüksek Öğretim Kurumu-YÖK), university rectors, public prosecutors, and police, as well as their own students (AfP-Germany, 2021; BAK, 2020; TELE1, 2020). There were attacks by nationalists, such as a pro-governmental mafia leader, who declared that he "will take a shower with the blood of the academics who signed the petition" (SCF, 2017).

Initially, there were 1,128 signatories; their numbers increased to 2,212 after the first attacks. Some of those working in Turkish universities were forced to resign, or their contracts were not extended. The state used the failed coup attempt in July 2016 as a pretext to dismiss the critical scholars by statutory decree laws (*kanun hükmünde kararname* [KHKs]), and blamed them for "terrorist propaganda." From July 2016 to July 2018, 125,000 people were dismissed by means of the KHKs, and around 5,000 were academics (Amnesty International, 2016; Erdem, 2018; Sade, 2020). Dismissal by KHK means that there will not be any job opportunities available in the public or private sectors and that the scholars' passports will be confiscated to prevent them from leaving the country. This was called "civilian death" (BAK, 2020; Gencel-Bek, 2018).

Some of these critical scholars left the county with short-term scholarships; some could not because their passports were confiscated. One-third of the signatories were PhD students; most were dismissed from their research assistant positions and some lost the professors who were guiding their theses. 822 critical scholars[2] faced criminal court hearings and they received penalties, were taken into custody, or were sent to prison.

There is no automatic recovery process for their return to their previous positions or compensation for the violations of their rights and freedoms, although the Constitutional Court (CC) has decided that "The punishment of academics due to signing the petition is the violation of freedom of expression" (Constitutional Court, 2019). Many of those dismissed scholars

---

2   A total of 763 were the first group and 59 were the second group of signatories; 108 received up to 36-month penalties, 12 were convicted, 96 were suspended, 719 were acquitted. There are 91 ongoing trials. Four academics in 2016 and a political scientist and mathematician in 2019 were imprisoned. A professor, an environmental activist, having pro-Kurdish *Halkların Demokratik Partisi* (HDP; Peoples Democratic Party) connections has been in prison since October 2020 (AfP, 2020; Amnesty International, 2016; BAK, 2020; TELE1, 2020).

by KHK – 406 people – appealed to the higher courts to return to their posts and get their vested rights (AfP, 2021). Most of these appeals were waiting for the decision by the OHAL Commission (*Olağanüstü Hal Komisyonu* [The State of Emergency Commission]) for around four years. The OHAL Commission refused some of these appeals at the end of October 2021, and there are many scholars still waiting for an answer. Scholars have to engage in lawsuits or appeals, which take time and necessitate financial resources, patience, solidarity, and support. By 2020, most of them had started the process to get new passports. However, those who had privileged green passports for working at public universities over 10 years could not renew them due to having been dismissed (AfP-Germany, 2021; BAK, 2020; Demir-Gürsel, 2017; HRFT, 2019).

Acar and Coşkan (2020, pp. 1–2) have discussed AfP's continuous engagement in scholarly activism, which has been driven by their "being dismissed via decree laws" and their reaction to "injustices." Their ideals about academia and academic freedoms that empower them "to demand for social change." Another study underlined AfP's continuing efforts to "think, debate, engage, teach and write" (Özdemir et al., 2019, p. 252).

Yet, these critical scholars find themselves in additional difficult circumstances once they go abroad. A scholar on a TV program during the fifth year of Peace Petition stated that their coming to Germany with temporary scholarships resulted in uncertainty about their stay abroad. Furthermore, there is uncertainty about their return to previous positions in Turkey (Özer, 2020). Özdemir and colleagues (2019) criticized academia as unsafe due to the neoliberal practices of high "competition and marketization" and "limited full-time positions" for professors. The competitive and scarce full-time positions generate more precarity among the scholars from abroad. To survive, they need to compete for third-party funding, leave academia for business or industry, or leave Germany.

This article argues that even though the critical scholars were dismissed in Turkey and had difficulties as a result of receiving only short-term scholarships in Germany, they established survival mechanisms such as founding solidarity academies and organized solidarity campaigns, and they supported each other during the court hearings guided by BAK-Litigation Coordination. They continue to criticize the increasing authoritarianism in Turkey and abroad and insist on their demand for peace. They have taken part in solidarity and critical networks and contributed to critical knowledge production during the conferences and workshops, considering topics such as precarious working conditions, refugee – migration, gender, and LGBQI+ issues as well as increasing nationalism and authoritarianism.

In light of the difficulties and their critical engagements, this article will concentrate on three questions: 1) What difficulties do critical scholars face in Germany; 2) What kind of survival and solidarity mechanisms do they develop; and 3) How do they change, challenge, or contribute to their host institutions or German academia?

Methodology

This research was based on 15 in-depth interviews with critical BAK scholars: 10 females and 5 males, and three hosts: one female and two males in Germany. Their ages ranged from 38 to 53 years. Nine of females and two of the males have children. Their career stages ranged from assistant professor to professor, and included one doctoral student. A total of 15 scholar and one host interview were conducted in Turkish, and two host interviews were conducted in English. Fourteen interviews were recorded from February through March 2020, and ranged from 25 to 100 minutes in length. The anonymized interlocutors are presented here as M (male) or F (female) plus age, for example, F1, 44. In addition, relevant parts of three interviews, one published in a newspaper, another streamed on a TV program by a male and two female academics (Özer, 2020), and the third interview was published in taz newspaper in Germany (Gökşin, 2020) and Bianet online in Turkey (Kural, 2017), have been used. The web page of the Academy in Exile (AiE) and the notes taken by the author in their conference were used as well (AiE, 2020).[3]

Results and discussion

The analyses will be structured according to the three main research topics: 1) difficulties, 2) solidarity and networks, and 3) challenges and contributions.

---

3   AiE was founded in 2017. It is a collaborative project of the Institute for Turkish Studies at the University of Duisburg-Essen, the Institute for Advanced Study in the Humanities (KWI), Essen, and the Forum Transregionale Studien, Berlin. It hosts 59 scholars for three to 24 months and one scholar for three years.

Difficulties

The difficulties can be clustered into three subtopics as follows:

Precarity - uncertainty - loss of control

The scholars who took part in this research had scholarships (or contracts) ranging from a few months to three years. A full professor who had come with her child and husband in 2017 explained the difficulties about short-term scholarships, bureaucratic systems, and passports:

> Two days after the dismissal, we came to Germany . . . the welcome center was closed. I lived with severe depression [in Germany] and had short 3- to 5-month scholarships . . . Then, I had a 21-month scholarship with a contract and another short-term one. Now, they offered me a 50% short contract . . . This is scary . . . I experienced precarity here . . . We have to fight many uncertainties. I did not have a passport [not renewed by Turkey, and no foreigner passport was given in Germany], and we could not travel because of a feeling about loss of control. (F9, 50)

Another scholar also explained the problem of not getting support from the welcome center during the summertime and later (F3, 45). A female associate professor stayed after the coup attempt, and she and her husband survived with short-term scholarships and contracts, but had a few unemployment periods lasting from 4–6 months (F6, 42). A male scholar, after being in two cities with only monthly scholarship in France, came to Germany with a two-year Philipp Schwartz Initiative (PSI) scholarship. He stated that the extension of PSI for another year has been possible since 2019. For this, he arranged a six-month scholarship through a third funding institution that was accepted by PSI, as the funding was not provided by his university (M2, 52). A host confirmed this saying that "when hosts have extra funding," they could contribute to the PSI's third year by adding six months (MHost1, 40).

Moreover, even if the scholarships are for two to three years, they are not designed to contribute to the social security system, which prevents the scholars from unemployment and retirement rights. In Germany, when individuals work continuously for 12 months with a contract, within 30 months, they are entitled to receive unemployment benefits (I am Expat, 2021). Many scholars have raised these issues, and PSI scholars have been offered contracts that contribute to the social security system since 2021. Thus, they are now entitled to the associated benefits and rights.

Nevertheless, the scholars must have long-term residence permits to receive unemployment benefits, which is not possible due to their limited contracts. Their residence permits end immediately or 15–30 days after the end of their contracts. This again terminates their legal rights to receive unemployment salaries. Moreover, the end of a contract determines the end of their housing contracts, placing them in a more precarious position than German scholars. There is the option to apply for a job seeker's visa; if they get it, they can apply for unemployment salaries (Germany Visa, 2021); however, it is an extremely difficult situation, as one female scholar explained:

> The health insurance we had did not even cover my children's vaccinations ... My son had eye surgery. We paid for it ... However, the feeling of insecurity is tremendous if something big happens, and we do not have decent insurance that would cover it. Plus, the moment we are unemployed, we will not get any unemployment salary, and we will not be able to retire at any stage of our life. We do not even have a visa [residence permit]. It expires when our contract ends ... The insecurity part is terrible. (F5, 49)

Having a contract enables the scholars to have public health insurance, which covers more health issues than scholars' inexpensive private insurance.

In addition, the limited number of full-time professorial positions in Germany contributes to a rising feeling of precarity among scholars. A male scholar stated that only the professors are not in precarious circumstances and even many German postdocs might never get a tenured professorship due to the high competition for full-time professorial positions. For this reason, one scholar noted, the critical scholars began to discuss leaving academia (M5, 40). A female host added that the German academic system is very different from other countries. According to her, there was "the urgent necessity" for these scholars to leave Turkey with short-term scholarships; some even came "to take a breath." Yet, the competitive system they have found in Germany makes it difficult for them to survive (FHost, 53).

In sum, most scholars rely on short- and fixed-term arrangements during which they need to finish and publish their research and apply for future competitive opportunities. They are also repeatedly faced with bureaucratic issues on residence permits, health insurance, rearrangement of schools or kindergartens, and house contracts after each short-term extension. Some did not even have valid passports until the beginning of 2020, which restricted their freedom of movement. They perceived precarity, insecurity, loss of control abroad, and insecurity about their return to previous positions, as was mentioned by another scholar in an interview. All these factors might force them to leave academia, but also might encourage them to collaborate with the locals to challenge the system together.

Inclusion vs. feeling like an appendix: Independent research institutes vs. universities

Another difficulty faced by the scholars is that of inclusion into the German academic system. Our interviews provided evidence that this also depends on the hosting institution. Some scholars offered more positive feedback about independent research institutes compared with universities. A male scholar who came to a research institute in the city center described this:

> There are two master programs in English, which makes it technically easy to collaborate with and adapt into the environment. It is an extraordinarily cooperative and facilitating environment, and provides a wide network from all over the world. (M1, 42)

A signatory couple came to Germany because of the attacks on signatories in other universities and the coup attempt in July. An independent research institute hosted the female scholar for a year. She described it as a collaborative international environment that encouraged her to organize a workshop for "at-risk scholars." She had a nice office and got support for accommodations, schooling, residence permit, and health insurance. After that, she became affiliated with a German university and described this change as "falling down":

> Our presence here [at the university] is perceived as temporary, like a person's appendix. We are supposed to be equal academics ... The host professor did not show any interest when I met him and explained my research and publications. Maybe he was too busy ... They see you as a person – even as a thing – who will occupy one of the rare office spaces. (F1, 44)

Another scholar shared her unfavorable experience with research institutes in two different universities: she had no office space in one and rare contact with the colleagues in the other (F4, 42). Others mentioned shared crowded office spaces or office space on a faraway campus (M3, 38; M4, 47).

Thus, the scholars made more favorable evaluations about independent research institutes, noting the benefits they had received. First, having a pleasant office space generated a collaborative environment and contact with colleagues. Second, attending or organizing workshops, seminars, and conferences with international scholars increased future collaborations. Third, easing the repetitive bureaucratic mechanism concerning accommodations, schools or kindergartens, residence permits or registrations, etc., was very helpful.

On the other hand, they criticized the universities or some research institutes in the universities for keeping them on the periphery, such as giving them a space in shared, crowded, or faraway campuses. This affected their concentration and limited their contact with other colleagues. Also, some busy professors, they noted, did not see them as valuable international scholars who had valuable experience but only as someone who took up limited space. Lastly, they reported that non-supportive welcome centers, especially during the summertime, made dealing with repetitive bureaucratic issues difficult.

(Re)-Habilitation and vested rights

A major problem has been the non-recognition of academic achievements acquired abroad in Germany and the demand to prove their scholarly experience. The interviewed scholars had requested their right to be recognized in Germany. In particular, redoing the "habilitation"[4] was questioned by all scholars. A senior scholar underlined resettlement of vested rights "to zero" in Germany:

> Our associate professorship title should be recognized as a habilitation . . . The preparation, the book, and then the exam takes one and a half years. Plus it's stressful. Then, when you came here, it was set to zero. (F8, 48)

Associated with this is the limited possibility to get a formal approval for supervising PhD students, as they were able to do in their previous positions in Turkey:

> For the moment, a PhD student asks me to be her supervisor . . . When the student requests, the university administration should decide. Still, there is no positive decision. . . . We started the process [to supervise the PhD

---

4   Habilitation is a qualification for professorship in Germany and Austria. To achieve this, a monograph or cumulative collection of articles, defense of the thesis, and giving a public lecture and course that shows teaching experience are necessary. See https://www.fu-berlin.de/en/sites/drs/postdocs/career/career_paths/habilitation/index.html. Similar requirements exist for the associate professorship in Turkey and other countries.

student] and my department is pushing for it. Some PhD students from Turkey started new PhDs here because their courses were not recognized. This is demotivating. However, one mentor put in a great deal effort to gain recognition of the courses of her PhD student, who started to do his thesis here immediately. (F8, 48)

The mentor who managed the get the recognition of the courses underlined that she even asked for an intervention of the international office for bureaucratic issues. But, she added, this requires extra effort and time, and many professors are very busy with their workload in Germany (FHost, 53).

An associate professor who was imprisoned after the Peace Petition came to a small town in Germany in 2017 with a 10-month scholarship, but then returned to Turkey after his PSI application was refused. One year later, he came to Berlin with a job arranged by SAR. He is now a PSI scholar. He worked on his habilitation during which time the university organized a private German teacher to support him. He stated:

It was a cumulative habilitation, and I wrote two articles . . . and delivered my habilitation thesis that passed the internal jury and outside reviewers. I gave a seminar and a lecture in German. There were three professors, one mitarbeiter [scientific employee], and a student representative on the defense committee. I graduated when my thesis was published in November 2019. (M5, 40)

These examples show that the associate professorship (and professorship) titles acquired in Turkey do not have value and were not accepted as equal to the habilitation in Germany. Yet, having a habilitation does not guarantee a permanent or professorial position, even for a German *privatdozent* (adjunct professor). Moreover, currently, they cannot supervise PhDs, but some are challenging the host institutions on this issue. Besides, recognition of PhD students' previous courses requires much effort and time by professors who have a heavy workload plus administrative tasks. Thus, Germany's competitive, hierarchical higher education system generates precarity for all scholars working with short-term scholarships or contracts, which is even more challenging for those scholars from Turkey or other non-European countries.

Solidarity networks and critical knowledge production

Since 2016, critical scholars in Turkey and abroad have engaged in solidarity and networking activities. They founded alternative centers for knowledge production, such as solidarity academies in the large cities of Ankara, Eskişehir, İzmir, and İstanbul, or smaller cities such as Dersim, or associations, cooperatives, cafes, and libraries like Kültürhane in Mersin. They produce critical knowledge in workshops, conferences, forums, or summer schools while bringing the signatories and their students together. One of these is Kocaeli Solidarity Academy (KODA), founded by dismissed scholars from Kocaeli University in 2016 (AfP-Germany, 2020; Georg-August-Universität Göttingen, 2017; KODA, 2021; Bianet, 2017b).

Over 200 critical scholars from Turkey took part in the AfP-Germany network. Most of them are members of the *Wissenschaftler*innen für den Frieden* (Academics for Peace–Germany Association) founded in Berlin in 2017 (AfP-Germany, 2021; Artı TV, 2021). Some are members of the online university platform and association, Off-University, that brings "dismissed scholars and their students" together in free online lectures and conferences (Off-University, 2017; Özer, 2020). Others are members of the Academics in Solidarity (AiS) based at the Freie Universität Berlin developing a peer-mentoring program for exiled and established scholars (AiS, 2019). They have contacts with female and migration networks such as Puduhepa or critical academic networks such as *Assoziation für kritische Gesellschaftsforschung* (Association for Critical Social Research) (AkG, 2018) and syndicates specialized in education such as *Gewerkschaft Erziehung und Wissenschaft* (Education and Science Union) (GEW, 2019). Many have contacts with at-risk institutes such as Scholars at Risk (SAR, n.d.). They are fellows of foundations including the Alexander von Humboldt Foundation (AvH) and its Philipp Schwartz Initiative (PSI), the Einstein Foundation, or AiE program (AiE, 2020; AvH, n.d.; Einstein Foundation, n.d.).

The critical scholars emphasized the solidarity and support among them. One mentioned receiving support from AfP-Germany for bureaucratic issues and arranging workshops (F4, 42). Another explained the help of networks connected to AfP-Germany, including lawyers, doctors, and schools (F6, 42). A third underlined the importance of having a network and institution (AfP-Germany) that provides the opportunity to "speak institutionally and politically" (F7, 39). A male scholar noted that colleagues from AfP-Germany have connections with left-leaning parties such as Die Linke and Greens or foundations like those of Rosa Luxemburg and Heinrich Boell, or trade unions and activists from the Kurdish movement (M3, 38). Another scholar, who returned to Turkey after receiving scholarships for five months from two institutions, came to Germany with an Einstein scholarship in 2021. He emphasized the support from the network:

> Following the trials, I applied to various scholarships in Germany with the help of scholars [from BAK]. My two friends here [one from BAK, one from Off-University] are like half of Berlin for me. I solved many problems by asking them for help. I also attended one day of Off-University's workshop in 2019. (M3, 38)

A full professor, who went to France with a short-term scholarship after being threatened by her students and then became a PSI fellow in Germany, said:

> I am a member of AfP-Germany, and active in the feminist – queer researcher network with some other signatories working on gender studies . . . I taught a seminar at Off-University. A PhD and a postdoc came to my city via my connections. I am a member of GEW and am working on my project about Academics in Germany. I helped GEW establish a lab on gender, and we will develop a network for PhDs working on gender issues. (F2, 53)

Many solidarity and support workshops were organized by and for AfP-Germany and Off-University networks, such as those on German and European research funding opportunities.[5] In some colloquia, the research proposals of PhD students and postdocs have been supported (F8, 48, F3, 45; Dressler, 2020). They gave free lectures and seminar series to Turkish-speaking communities, as the one in Cemevi Berlin (Berlin Alevite Community) in 2018 (Cemevi, 2018). After the pandemic, the AfP-Germany network transformed its gathering designed to exchange information and discuss specific topics (also cook together) in *Mahalle* (Neighborhood) into an online forum called *Sanal Mahalle* (Digital Neighborhood). They discussed topics such as gender, state politics or COVID etc (AfP-Germany, 2020). In addition, they organized a solidarity campaign for their colleagues in Turkey (Önaldı, 2017), and a new fund-raising campaign on "The academic solidarity across borders" was initiated (ASaB, 2021). Moreover, the AfP members in

5   Such as the author's contribution in the Off-University project workshop with Dr. J. Strutz and B. Bayraktar in Bielefeld, June 8–10, 2018 or the BAK-Off project workshop, Berlin, December 6–7, 2018 with Dr. S. Kirmse.

Turkey were calling the signatories by phone to check their current situation after a newly graduated signatory committed suicide in February 2017 who left to "civilian death" without any job or fellowships (Bianet, 2017a). They either called or sent emails to the approximately 700 PhD students or research assistants who had originally signed the petition, asking them to update their academic, financial, or psychological situation. This solidarity responsibility was inherited by a group from AfP-Germany in 2019. They also supported and guided them to voluntary mentors and psychologists via established networks. They shared calls for scholarships or conferences, guided them in their applications for scholarships, and shared information related to bureaucratic or administrative issues. The members distributed information via their listserve *Doktora öğrencisi ve araştırma asistanı destek grubu* (DAD, PhD student and research assistant support group), solidarity academies, and people on the AfP network.

Challenges and contributions

The interviewed critical scholars cited examples of how they challenged the host and funding institutions, which resulted in additional opportunities within the system such as the extension of PSI scholarship for another year or new opportunities like the AiE program (AiE, 2020). They also talked about their contribution to German academia on critical topics and their publications. Some criticized the local scholars for their acceptance of the neoliberal and competitive system leading to precarity for PhDs and postdocs "as a given" without doing anything (F8, 48).

On the other hand, one scholar said their challenges were small, and bigger challenges could be possible when they continuously collaborated with the critical scholars in Germany, such as when they opened several panels at the AkG conference in 2018 (F7, 39).

The interlocutors involved in this research mentioned that the AvH foundation has been one of the most collaborative institutions because it listens and takes notes about scholars' challenges and tries to provide solutions. One example is the extension of PSI fellowships for "6+6 months," which was brought to the agenda by the scholars during the conferences, and increasing the lump sum given to the host institutions in 2019 (AvH, n.d.). One critical scholar underlined their active involvement during a conference in 2018. She added that the scholars who constituted this new intellectual migration had made contributions with their critical approaches:

We [AfP-Germany] were prepared well and presented the challenges about the scholarships, residence permits, or health insurances, and PSI, SAR, and our hosts listened to us. A German scholar has made a very critical speech [about the problems we face here]. There is no discussion. We have influence ... and we attract much attention within the new wave of intellectual migration, which is not only white-collar but also golden collar ... Many products [publications] will come out in a few years ... When I first talked about the German education system's insensitivity to neoliberalism, even my mentor was resisting. Now, we [together] are using this critical approach in our research. (F2, 53)

A male scholar said that they have an impact, but it is "within the priorities and policies" of Germany's donor organizations and educational institutions (M1, 42). A law scholar advocated for a one-year extension of PSI fellowships for pregnant scholars, which she underlined as "a legal right in Germany" (F8, 48). The issue raised here pertains to the problems faced by PSI scholars who need to present their requests for fellowships, convince the funding organization of their eligibility, and request the support mechanisms that exist for other fellowships of the same institution (AvH) and which are often not considered for them. For example, a female scholar with children who receives a George Foster scholarship awarded by AvH has the right to request a one-year extension. Moreover, other AvH scholarship holders have the right to take intensive German courses before the start of the fellowship, making their adaptation and resettlement process easier, while PSI scholars have to cover their language courses from the lump sum provided to their hosts when receiving their fellowship. Indeed, it is not easy to do everything at one time: resettle, register, find schools for children, start German courses, conduct their research, publish, and apply for future funding in two (or three) years. In addition, all previously funded AvH scholars become alumni and have the right to request short-term scholarships in the future, but this is not yet the case for PSI scholars.

A host and a signatory to the petition stated that they challenged the Einstein Foundation's policy which is based on one host–one scholar by applying on behalf of eight critical scholars from two departments (MHost2, 45).

Another scholar made this point:

We knew to be critical wherever we have been ... Once, I told the previous foreign minister that "the system in Germany will never include us, and we will continue to be precarious." The Chairman of Humboldt [Foundation] strongly opposed me. Many scholars are doing research on precarity now (F10, 45)

One host noted their support for the scholars' contributions: "I am happy with new critical topics and discussions toward the neoliberal university system raised by AfP. I am not sure how long it will continue. . . . It is essential to have partners in critical discussions" (FHost, 53).

Most interlocutors mentioned that they gave lectures in English, even if it was not required of them as part of their teaching. They explained that they were searching for ways to supervise PhD students and were serving as second supervisors for master's students. They also actively organized collaborative colloquia for the students (F2, 53; F3, 45; F5, 40; F6, 42; F7, 39; F8, 48; F9, 50; M1, 42).

In sum, these scholars are contributing and continue to contribute to the German academic system with their publications and critical perspectives that have opened up new discussions. This underlines the importance of having highly qualified scholars from Turkey included in the German system who will contribute more in the long term.

Concluding remarks: From guest worker to guest researcher

The research findings suggest that critical scholars from Turkey face difficulties due to short-term scholarships, non-recognition of vested rights (associate professorship) and previously taken courses (for PhDs). There is a foreign bureaucratic system which is difficult to navigate with limited German language capabilities, and they are being kept on the periphery of academia in some institutions. Some have not been able to obtain new passports, therefore their freedom of movement is restricted.

Their positions are more precarious than those of native Germans. The problems in connection with short-term scholarships also affect social security, retirement and unemployment rights, health insurance, and bureaucratic issues. Yet, because around 90% of scholars working in academia have positions lower than professorships with limited contracts or fellowships, they all work in the same neoliberal and competitive academic system based on "marketization and hierarchization" (Özdemir et al., 2019, p. 252). Therefore, a structural change in favor of academia necessitates the collaboration of all scholars in Germany. The scholars agree that they have to challenge and influence the host and funding institutions through their collective or private efforts. Their efforts can be more effective if they organize (F2, 53) and collaborate with critical German (F7, 39) and international scholars. It is too early to say if the aforementioned solidarity initiatives will have an effect on German academia, but it is certainly worth observing how these initiatives affect other initiatives and debates in a system where it is recognized that precarity has become the rule rather than the exception.

However, it must be noted that the German academic system offers an exit point to the critical scholars, similar to the local PhDs and postdocs, to leave academia in favor of industry or opening their start-up businesses instead of waiting for professorial positions. This issue has been discussed during the conferences organized by PSI (AvH), risk institutions, or solidarity programs such as AiS since 2019 (M5, 40; see AiS, 2020).

Despite these difficulties, the scholars continue to develop solidarity mechanisms and establish associations or solidarity academies in Turkey and abroad. They help their colleagues in Turkey with their applications abroad and establish contact with possible mentors. Moreover, a group of scholars from AfP-Germany continues to follow the signatories in need and organizes solidarity campaigns.

Critical scholars challenge the funding and host institutions to develop cooperative mechanisms and create additional opportunities while pushing the boundaries. In addition, they organize workshops and conferences and publish articles and books, the number of which is increasing. They give lectures and seminars in English and contribute to their institutions' internationalization (F2, 53; F8, 48; M1, 42; see Özer, 2020).

Moreover, they contribute to critical knowledge production and open new topics on precarity and the neoliberal system in German academia, which is perceived positively by German scholars, like the host professor interviewed in this research (FHost, 53). They criticize the authoritarianism and increasing nationalism in Turkey and other countries. They engage in political activism such as organizing campaigns or demonstrations about the imprisonment of their colleagues, such as for Prof. Dr. Üstünel or Assoc. Prof. Altınel in 2019, or illegitimate assignment of rectors, such as the case at Bosphorus University (Yalaf, 2021). They insist on demanding peace in Turkey and abroad and fight against injustices wherever they are (Acar & Coşkan, 2020).

Germany is a country of migration, not only "becoming a country of migration," as Chancellor Merkel said (DW, 2015). However, the policies toward immigrants show a short-term perspective. In the 1960s, "guest workers" were supposed to leave two or three years after they came from Turkey. Now, intellectual, "golden-collar" immigrant guest researchers have become more common in Germany. However, if uncertainty, precarity, and keeping critical scholars on the edge of academia continue in Germany (F1, 44; F4, 47; M1, 42; see Özer, 2020), a small sign of normalization and democratization will make their return to Turkey quicker. Germany had a similar experience when highly qualified German citizens with immigrant backgrounds returned to Turkey after the EU negotiations resulted in a democratization and peace process that lasted until 2013, since then the pressures increased in the country. In 2010, the "negative migration balance of Germany vis-à-vis Turkey was 5,862 individuals" (Focus Migration, 2012, p. 1).

The following statement by a female scholar who returned to Turkey a few months after this interview explains the current conditions and the decision involved in whether to stay or leave:

> To stay here, I need to have peace with what academia means in Germany: to talk and think in a more isolated environment, design your life in accordance with future projects, and concentrate on bringing new funding . . . We are at a border between entering into it and not. My brain is in a state of being in limbo. If I stay [in Germany], I need to concentrate on my big project. If I decide to return, I need to concentrate on my associate professorship. In Turkey, I have to think about the earthquake and the war. It is uncomfortable to think about returning to Turkey when you have a child while living in peace and prosperity here. On the one hand, we had a friendly environment [Turkey] where I always felt comfortable. I used to be something here also, but we all know what I am here [an "appendix"]. I remember my students [there] . . . They were the most essential and rare elements left to us. Although I was very bored with the teaching load . . . our relations with students were meaningful . . . Despite some students spying on us . . . I still hope that we could impress even the AKP supporter students . . . I miss our relationship with students . . . I am still in between staying or leaving. (F, 44)

References

Academic Solidarity across Borders – ASaB (2021). Academic Solidarity across Borders. Retrieved from https://www.academicsforpeace-germany.org/s olifund/

Academics for Peace – AfP (2020). Three years of rights violations faced by Academics for Peace. Hypotheses. Retrieved from https://afp.hypotheses .org/1482

Academics for Peace – AfP (2021). Rights Violations Against Academics for Peace. Retrieved from https://barisicinakademisyenler.net/node/314

Academics for Peace-Germany – AfP-Germany (2020). Sanal mahalle toplantıları basliyor [The digital neigbourhood meetings start]. Retrieved from https://academicsforpeace-germany.org/2020/05/06/sanal-mahalle -toplantilari-basliyor/

Academics for Peace-Germany – AfP-Germany (2021). Home. Retrieved from https://www.academicsforpeace-germany.org/

Academics in Solidarity – AiS (2019). Academics in Solidarity. Retrieved from https://www.fu-berlin.de/en/sites/academicsinsolidarity/index.html

Academics in Solidarity – AiS (2020). Webinar series 'Research funding'. Retrieved from https://www.fu-berlin.de/en/sites/academicsinsolidarity/ news/WebS-Research-Funds.html

Academy in Exile – AiE (2020). Academy in Exile- Fellowship fact sheet. Retrieved from https://www.academy-in-exile.eu/fellows/fellowship-fact -sheet/

Acar, Y. G., & Coşkan, C. (2020). Academic activism and its impact on individual level mobilization, sources of learning, and the future of academia in Turkey. *Journal of Community and Applied Social Psychology (Online)*, 1–17. DOI:10.1002/casp.2455

Alexander von Humboldt Foundation – AvH (n.d.). AvH- Programmes A to Z. Retrieved from https://www.humboldt-foundation.de/en/apply/sponsors hip-programmes/programmes-a-to-z

Amnesty International (2016). Urgent Action: Academics detained for signing peace appeal. Retrieved from https://www.amnesty.org/download/Docu ments/EUR4437922016ENGLISH.pdf

Artı TV (2021). Ezo Özer ile- Odak Barış için Akademisyenler bildirisinin üzerinden 5 yıl geçti [It has been 5 years since the Academicians for Peace's Petition] [Video]. *Youtube*. Retrieved from https://www.youtube.com/wat ch?v=KV8L-mUq_RI

Assoziation für kritische Gesellschaftsforschung – AkG (2018). Antidemokratische Konservative/Tagung 2018. Retrieved from https ://akg-online.org/tagungen/20node8-antidemokratische-konservative-t agung-30nodenode-2node22onode8

Barisicin Akademisyenler – BAK (2020). Reports. Retrieved from https://bari sicinakademisyenler.net/node/4

Bianet (2017a, February 25). Eğitim Sen: Akademisyen Mehmet Fatih Traş intihar etti [Academician Fatih Traş had generated suicide]. Retrieved from https://bianet.org/bianet/toplum/183975-egitim-sen-akademisyen-mehmet-fatih-tras-intihar-etti

Bianet (2017b, October 3). Solidarity academies to start new academic year. Retrieved from https://bianet.org/english/human-rights/190281-solidarit y-academies-to-start-new-academic-year

Cemevi (2018). BAK- Almanya: Türkiye Seminerleri [Afp-Germany: Seminars on Turkey, Alevitische Gemeinde zu Berlin e. V.]. *Facebook*. Retrieved from https://www.facebook.com/events/berlin-alevi-toplumu-cemevi-al evitische-gemeinde-zu-berlin/bak-almanya-dayan%C4%B1%C5%9Fma-a kademisi-t%C3%BCrkiye-seminerleri-1/166430997325119/

Constitutional Court (2019). Zübeyde Füsun Üstel ve diğerleri başvurusu [Appeals by Zübeyde Füsul Üstel and others, Nr. 2018/17635, dd. 26/7/2019; Official Newspaper dd.19/9/2019, Nr:30893]. Retrieved from https://karar larbilgibankasi.anayasa.gov.tr/BB/2018/17635

Demir-Gürsel, E. (2017, December 11). Esra Demir-Gürsel: A brief note on the "crime" allegedly committed by the Academics for Peace: Propaganda for a terrorist organization or degrading the State of Turkish Republic? *Hypotheses*. Retrieved from https://afp.hypotheses.org/236

Deutsche Welle – DW (2015). Merkel: Germany is becoming a 'country of immigration'. Retrieved from https://www.dw.com/en/merkel-germany -is-becoming-a-country-of-immigration/a-18491165

Dressler, M. (2020, May 15). Networking and Mentoring Workshop in the Humanities and Social Sciences. Online mini-conference mentoring and networking for PhD students at risk. Retrieved from https://www.multi ple-secularities.de/events/event/networking-and-mentoring-workshop-i n-the-humanities-and-social-sciences

Einstein Foundation (n.d.). Einstein Foundation. Retrieved from https://ww w.einsteinfoundation.de/index.php?id=1&L=1

Erdem, O. (2018, July 17). OHAL sona erdi: İki yıllık sürecin bilançosu [The state of emergency ended: The balance sheet for two years]. *BBC Türkçe*. Retrieved from https://www.bbc.com/turkce/haberler-turkiye-44799489

Focus Migration (2012). From home country to home country? The emigration of highly qualified German citizens of Turkish descent to Turkey. *Focus Migration*, 17, 1–8.

Gencel-Bek, M. (2018). Life, journey, migration: Enforced mobilization of an academic. *Movements. Journal for critical migration and border regime studies*, 4(2), 225–231.

Georg-August-Universität Göttingen (2017, February 10). Politically engaged academia: Solidarity with the Academics for Peace in Turkey and Germany. Retrieved from https://www.uni-goettingen.de/de/konferenzen/357000.html

Germany Visa (2021). Germany: Job seeker visa. Retrieved from https://www.germany-visa.org/job-seeker-visa/

Gewerkschaft Erziehung und Wissenschaft – GEW (2019, May 24&25). Finding common solutions: Precarious labour and the lack of diversity in German higher education. Retrieved from https://www.gew.de/veranstaltungen/detailseite/finding-common-solutions-precarious-labour-and-the-lack-of-diversity-in-german-higher-education/

Gökşin, O. D (2020, June 6). Göçmen akademisyenler anlatıyor: 'Almanya'da profesör Allah gibi' [Immigrant Academics Tell: 'The professor is like God in Germany']. *taz – die tageszeitung*. Retrieved from https://taz.de/Goecmen-akademisyenler-anlatyor/!5665668/

Human Rights Foundation Turkey – HRFT (2019). Academics for Peace: A brief history. Retrieved from http://www.tihvakademi.org/wp-content/uploads/2019/03/AcademicsforPeace-ABriefHistory.pdf

I am Expat (2021). Unemployment benefits in Germany (Arbeitslosengeld). Retrieved from https://www.iamexpat.de/expat-info/social-security/unemployment-benefits-germany-arbeitslosengeld

Kocaeli Solidarity Academy – KODA (2021). Bahar dönemi: Atölyeler/tartışmalar [Spring semester: Workshops /discussions]. Retrieved from https://www.kocaelidayanisma.org/kategori/egitimler/

Kural, B. (2017, December 19). Dört akademisyenin Berlin'i – Kılvılcım: İki ülkede de üniversiteler aynı sistemin baskısı altında [Berlin of four academics- Kıvılcım: Universities in both countries are under the pressure of the same system]. *Bianet*. Retrieved from https://bianet.org/b

ianet/ifade-ozgurlugu/192479-kivilcim-iki-ulkede-de-universiteler-ayni-sistemin-baskisi-altinda

Off-University (2017). About us. Retrieved from https://off-university.com/en-US/page/about-us

Önaldı, O. (2017, September 10). Cemevi barış akademisyenlerine dayanışma gecesi düzenledi. [Cemevi has organized a solidarity night for the Academics for Peace]. *Haber*. Retrieved from https://www.ha-ber.com/cemevi-baris-akademisyenlerine-dayanisma-gecesi-duzenledi/90819/

Özdemir, S. S., Mutluer, N., & Özyürek, E. (2019). Exile and plurality in neoliberal times: Turkey's Academics for Peace. *Public Culture, 31*(2), 235–259.

Özer, T. (2020, November 23). Barış İçin Akademisyenler: Almanya'da güvencesizlik, Türkiye'de işsizlik [Academics for Peace: Insecurity (Job) in Germany, unemployment in Turkey]. *Cumhuriyet*. Retrieved from https://www.cumhuriyet.com.tr/haber/baris-icin-akademisyenler-almanyada-guvencesizlik-turkiyede-issizlik-1793047

Sade, G. (2020, July 15). 15 Temmuz darbe girişimi sonrasında kaç kişi görevinden ihraç edildi, kaç kişi tutuklandı? [How many people were dismissed and arrested after the coup on 15 July]. *Euronews*. Retrieved from https://tr.euronews.com/2020/07/15/verilerle-15-temmuz-sonras-v e-ohal-sureci

Scholars at Risk – SAR (n.d.). Scholars at Risk. Retrieved from https://www.scholarsatrisk.org/

Stockholm Center For Freedom – SCF (2017, October 25). Pro-Erdoğan Turkish mafia leader defends threat against Academics for Peace. Retrieved from https://stockholmcf.org/pro-erdogan-turkish-mafia-lead er-defends-threat-against-academics-for-peace/

TELE1 (2020). 20 HDP'li siyasetçiden 17'si tutuklandı, Kars Belediyesi'ne kayyum atandı [17 out of 20 politicians of HDP have been arrested: A trustee has been assigned to the municipality of Kars]. Retrieved from https://tele1.com.tr/adliyede-hdplilere-kalkanli-polis-mudahalesi-235296/

Yalaf, E. (2021, January 9). Berlin'de Boğaziçi'ne destek gösterisi [Support protest for Bophorus University in Berlin]. *Deutsche Welle*. Retrieved from https://www.dw.com/tr/berlinde-bo%C4%9Fazi%C3%A7ine-destek-g%C3%B6sterisi/a-56181758.

Networks matter[1]
Scholars in exile in Germany and their relations to home and host communities

Carola Richter

According to the British Council for At-Risk Academics (CARA, 2017, p. 6), the years 2016–2017 marked the highest number of at-risk academic applicants since the 1930s, indicating strong pressure on academia on the whole. Scholars from war-torn countries in the Middle East and North Africa (MENA) region, such as Syria and Yemen, felt pressured to flee. Researchers from Turkey after Erdoğan's heavy suppression of dissent (Aktas et al., 2018) also fled. Hundreds of academics went into exile, leaving their home country for indeterminate periods. It is assumed that being forced into exile affects the nature of social networks in many ways, creating new dependencies – but, possibly, also new opportunities.

This chapter explores how professional networks of academics in exile are constituted. Who do they consider important to advancing their career? With whom or what are these persons or institutions associated – the current host country, the home country that they left, or the transnational sphere? Further, the study examines the communication modes being used to connect with various people in the network. Can pervasive online communication help sustain transnational networks and overcome the rupture caused by exile?

I discuss the answers to these research questions herein, applying a qualitative network analysis (QNA) of fourteen academics in exile from Turkey and the Middle East who are now located in Germany. Some findings in this study on academics in exile correspond to those of the problems of other mobile academics (e.g., unsatisfying conditions at the home institution, lack

1 An earlier version of this chapter has been published in the Journal of Global Diaspora & Media, 2(1) in 2021. The author would like to thank several Master students from Freie Universität Berlin who helped conducting the interviews.

of job security abroad, and dependence on a host). It can be argued, however, that being in exile adds the factor that scholars are "only conditionally accepted" (Kettler, 2011, p. 206), which comes with the "everyday concerns of asylum, livelihood, and isolation," "practical relations to the play of power," and "emotional stress" (p. 208) in the host society. Moreover, exiled academics often have "unfinished business with those they are compelled to leave behind" in their home nation, and the "consuming question of return" (p. 208) always looms.

A suitable analysis of academics in exile thus requires a broader context of the meaning of the restraints to academic freedom at both the individual and network levels. It also needs to consider how exile is connected to asymmetric power relationships and the resulting personal perception of belonging. In the next section, I demonstrate how this impacts networks.

The importance of networks for academics

Manuel Castells has shown in his seminal work, *The Rise of the Network Society* (1996), that networks are pervasive. They have become even more important due to technological advances of the 1990s that facilitated links with others through the internet. In an academic system, the literature often distinguishes between formal and informal networks (Kreis et al., 2019, p. 302). Formal networks refer to networks developed through membership in a university, department, or professional association. Informal networks, on the other hand, are based on personal relationships. Informal academic networks often originate from previously formal contexts, such as conferences or research collaborations, and then develop more autonomously in terms of personnel and content. This means that formal and informal networks are strongly interdependent: an effect in one part of a formal network relationship could affect one's informal network downstream. At the same time, a specific position or relationship within an informal network may also help one's status within a formal network.

Although traditional power centers, such as the state, can be bypassed by using technology to form networks, networks are not free from asymmetric power relationships. In his *A Network Theory of Power*, Castells (2011) speaks of various aspects of power, including "network-making power" as "the power to program specific networks according to the interests and values of the programmers, and the power to switch different networks following the strategic alliances between the dominant actors of various networks" (p. 773). In the case of scholars who have been cut off from their formal networks due to war or political pressure, several institutions have been established to assist in network-making in times of risk and exile. This development began in the 1930s during the mass expulsion of scholars from Germany and Nazi-occupied territories in Europe. Britain, the US, and Switzerland created committees to aid emigration from risky places. They assisted in finding host institutions for threatened scholars and familiarized them with the different academic cultures (Löhr, 2014, pp. 231–232). Since then, many institutions have evolved, such as CARA, the US-based Scholars at Risk (SAR), and the Scholar Rescue Fund (SRF), which aim to provide support to displaced scholars at North American and European universities and support them and their hosts through training and networking events. They work together with universities or mostly Western national funding organizations, such as the German Philipp Schwartz Initiative (PSI) of the Alexander von Humboldt Foundation (AvH). For example, from 2014 to 2019, PSI granted 196 fellowships to at-risk scholars in Germany. Of those, 116 went to scholars from Turkey alone (Brandstädter, 2019).

While these institutions are indeed important in lending scholars a helping hand at a crucial time of rupture in their networks, they are not without an "agenda," as Löhr (2014, p. 235) has put it. She argued that building these institutions in the 1930s was meant to strengthen the respective host university systems in a "dynamic interplay between the nationalization of educational systems and the simultaneous transnationalization of scientific communities" (p. 236). However, the intellectual transnational networks established could be strategically exploited to enhance reputation and gain support for national Western liberal systems of higher education and by referring to a national brain gain: "Help for refugee scholars meant to benefit from their talents" Löhr concluded (2014, p. 240). From a critical political-economy point of view, Cantwell (2011) argued that in this framework of "gaining brains," still today "the higher education sector, as a whole, and individual universities are viewed as instruments to attract and retain skilled knowledge workers around the globe in public policy designed to enhance economic productivity and competitiveness" (p. 428). Despite ostensibly "saving" scholars and providing them opportunities in exile, the described institutions are also part of what Fahey and Kenway (2010, p. 571) termed the "privileged receiving nation-states" within the "global geographies of power/knowledge."

Thus, network-making power is definitively related to asymmetric power relations within which exiled scholars have to maneuver. Research on networks is also about inclusion and exclusion, opportunities and dependencies, and their related power dynamics. This includes a focus on both spatial and communicative dimensions. Regarding the spatial dimension, Vertovec (2002) noted that networks of skilled workers – to which scholars belong – are ultimately transnational in character. Indeed, transnational mobility is a requirement for scholars to foster collaborations in multinational consortia, disseminate research findings in conferences and workshops, or engage in prestigious fellowships and fieldwork abroad. Thus, from an early point in their careers, scholars are encouraged to build networks beyond their national borders. However, in the case of exiled academics, these networks are formed in the framework of asymmetric power relations between Western liberal systems and more authoritarian-controlled systems. The former are also characterized by stronger economic (funding) power. The degree of autonomy of academic systems and related funding opportunities in the different countries also strongly affect network formation. Thus, in this study, transnational relations and the relations to both the host country and the home country were analyzed.

Social networks cannot exist without communication. Hepp et al. (2016) highlighted the relevance of communication in networks and the role of "polymedia": the various kinds of media and communication providing network formation and maintenance extending from digital media communication to classical letters (p. 88). Networks, indeed, "are marked by patterns of communication or exchange of resources and information along with participation in sociocultural and political activities" (Vertovec, 2010, p. 573). The modes of communication, however, reflect power dynamics, as some modes might be more or less accessible, more or less controlled, or considered more or less relevant to different individuals.

An analysis of scholarly networks in exile is of utmost importance in adequately assessing the power relations in which the exiled scholars are located and perceive themselves. Both spatial and communicative dimensions need to be explored to more effectively understand scholars' networks.

Methodology: A qualitative network analysis

I used QNA to investigate the professional network relations of scholars in exile and the communication modes they employ to form and uphold these relations. To date, few attempts have been made to utilize work with QNA on skilled migrants (see Bilecen & Amelina, 2017; Scheibelhofer, 2006).

As opposed to QNA, social network analysis (SNA) is more common in current research. While large networks can be reconstructed through SNA, the aim of QNA is different. Ahrens (2018, p. 1) explains that QNAs "apply a micro-perspective" and "prompt a focus on personal and social networks, for instance, friendships, family relations, or migration ties." QNA is thus a "useful technique to shed light on personal perceptions of reality and relationships individuals develop" (p. 3). QNA, according to Hepp et al. (2016, p. 89), refers to "[...] those forms of network research that render network structures and practices visible."

In QNA, personal networks are termed egocentric networks (Hollstein, 2016). In the literal center of such a network stands the person of interest, the ego, and around this ego, several alteri are positioned. To reconstruct the network as perceived by the ego, various techniques can be used, but the main instrument is an interview with the ego, or person.

I opted for QNA using interviews to obtain the data, including name generators and network cards. First, I developed an interview guideline that contained three major parts. The first part gathered information on the biography of the exiled scholar, asking, for example, about former and current job positions. The second and main part of the interview opened with the key question: "In a person's academic career several people and institutions are relevant to progress. Who is relevant in your professional network?" In order to stimulate the interviewee's memory, a name generator (Burt, 1984, p. 296) was used. A name generator is an empty list in which the interviewee is asked to put names, acronyms, or even numbers. In my case, this included persons or institutions that were considered relevant to the interviewee's professional career. In the next step, I asked the exiled scholar to order these names on a network card of concentric circles, explaining the relation to this person or institution. The concentric circles with the ego in their center symbolized the relative closeness of an alter to the ego as perceived by the ego. The network cards functioned as cognitive help in the interview to compare the position of several alteri; it also helped to stimulate discussion (Hollstein, 2016). In our case, we asked the interviewees to locate their identified alteri in one of three geographically defined sectors: the host country (i.e., Germany), in the home country (e.g., Turkey), or in the transnational sphere (i.e., anywhere except the host or home country). We asked the interviewees to explain their choice by locating a person in the different circles and to define the relationship with that person or institution in more detail, instigating this with the network cards. Furthermore, we asked the interviewees about their means of communication between the ego and the alteri. They were probed on the intensity of the communication and asked to reflect that by drawing thinner or thicker lines on the network card. Finally, the third part of the interview was a personal evaluation of the interviewee's current situation in Germany and visions of the future.

Interviews were done with the help of master's students from Freie Universität Berlin. We conducted interviews with 14 scholars in exile coming from different disciplines – mainly from the social sciences and humanities – from August to November 2019. Most interviews were conducted in English, some in German, depending on the scholars' preferences. The scholars were located in Germany – most in Berlin, a large metropolitan city with three universities that host a large number of scholars at risk in Germany. All interviews were conducted face-to-face, often in the offices of the interviewees, and lasted around an hour. A total of 12 scholars were originally from Turkey, one from Syria, and one from Yemen. Nine of the scholars were female and five were male. Their ages ranged from the mid-30s to early 60s, with most being from 40 to 50 years of age. Neither the country of origin, the disciplines, age group, nor gender of the interviewed scholars are representative. However, the sample reflects certain trends with regard to those scholars who have come to Germany. We also applied two conditions to allow a more homogeneous sample. The interviewee should have reached a senior or mid-career phase (hold a PhD and have held a position in academia in their country of origin) and have been exiled to Germany within the previous two to five years.

To find subjects for the QNA, we looked in the media reports for people who self-identified as scholars at risk, asked German scholars who worked as mentors to these academics, and went to SAR conferences. The potential interviewees were approached in various ways, for example, through personal contacts via the snowball method, by using email lists, or by contacting them at conferences. Although complete anonymity of the interviewee, including that of the person's alteri, was guaranteed, the response rate was rather low, and several attempts had to be made to convince interviewees to participate.

Most likely, the time-consuming, in-depth method and the possible revelation of sensitive personal information were major reasons for not taking part in the study. In addition, the overall population investigated in this study was small – only a couple hundred people – so 14 interviews were a good result for an exploratory study.

All interviews were audio-recorded and condensed in written "thick descriptions" (Geertz, 1973). These descriptions highlighted the main findings, which were supplemented by final network cards containing the sectoralized relationships of the ego to the alteri. The analysis relied on these fourteen network cards and the comparative qualitative content analysis of the 14 interviews.

Presenting the networks of scholars in exile

In the following sections, I present the networks of the interviewed scholars as snapshots in time. Therefore, the respective context will be presented before I delve into a reconstruction of the networks regarding the spatial (i.e., sectoral dimension) followed by the communicative dimension.

The networks' contexts

All interviewees had originally been well settled in their home countries, and all but one had previously attained a permanent professorship that they had permanently left or that was on hold with only a minimal chance of getting back upon return. All the Turkish scholars had signed the Academics for Peace petition (see Özgür in this volume), except one who had not supported it. One had been imprisoned after signing the petition, another was in pre-trial detention, and many had faced court trials before they went into exile. The interviewed Turkish scholars had mainly fled for political reasons, while the two Arab scholars did not provide any details about the reasons for fleeing their home countries. The scholar from Yemen believed that the regime was monitoring scholars ("the walls have ears," Interviewee 13) and believed that he was forced into retirement because of his critical remarks about the regime and had no chance to continue in academia in Yemen.

All interviewees had a temporary fellowship – most of them from the PSI – and some had already passed the first phase of PSI and were now in the second round of temporary fellowships awarded by institutions, such as the Berlin-based Einstein Foundation. All of these fellowships are conditional upon a German host institution. This means a university or institute and a hosting professor at this facility are needed to obtain the grant, which places the fellow in a kind of dependent post-doc position. The funding is then channeled through this institution, thus providing a financial incentive for the host university to participate. The fellowship is in most cases for only two years, sometimes with the possibility of extending it to three years.

Having given up permanent (professorship) positions in their home countries and now being faced with a series of temporary contracts (on more of a post-doc level), all of the interviewed scholars stated that the most complicated aspect of their professional career in Germany was the uncertainty of their academic future. The scholars were explicit about that. One said, "I do not make any plans anymore. It is about survival now" (Interviewee 6); another noted, "I do not see myself as a professor here. I gave up this thought. I am exhausted" (Interviewee 12). All interviewees subscribed to a mentality similar to that of one interviewee who stated, "Because we spend most of the time applying for another position or project . . . this project type of working creates much stress and is not very productive. . . . I feel that we are like the guest workers in the 1960s although we are intellectuals" (Interviewee 1). This job insecurity was also apparently connected to a loss of reputation that many of the scholars felt: "I am just a post-doc now" (Interviewee 4).

Another interviewee added an insight into the complexity of their situation: "We are coming as academics, but we are moving here with the family. The academic part is only one part of the struggle. The other private part is the bigger one" (Interviewee 3). This emphasizes that a lot of time that could be spent on networking was actually spent on family affairs. Those from Yemen and Syria, and the Turkish scholars with an asylum status, named travel restrictions as harmful to their networks, as they were not allowed to travel beyond the 26 Schengen Area countries.

Notably, most of the scholars interviewed had previously been to Germany for a longer period of time or had other connections to it. Many years before, two had done their early studies in Germany, and another interviewee had a partner with German citizenship. For them, these aspects were impetuses to come to Germany in their difficult situation, but did not have an immediate effect on their professional network, since contacts from these former times were based on private, non-academic factors. One of those scholars mentioned that he had written "more than 50 emails" (Interviewee 14) to

German professors asking them to be his host until he ultimately found one. He stated that it was only after arrival that "I reactivated my German network."

Others had completed their PhD work in Germany or had spent substantial time on scholarship in Germany during their nascent academic career. Their contacts with German academia had developed during these periods. These proved crucial to their professional careers and were of benefit during the tumultuous situation they had found themselves in before flight. Some interviewees mentioned that they actively contacted a former mentor, supervisor, or colleague at a German university for help in finding a host for a scholarship program (Interviewees 9, 12). The interviewee from Yemen had come almost every year to Germany for short-term visits and had made a lot of contacts "that were like bricks on which I could build a base now" (Interviewee 13). Others, like the interviewee from Syria, who had no previous experience with the German academic system and no established personal contacts, felt lost and had more difficulty settling into the new situation. In his case, SAR brought him into contact with possible hosting institutions and facilitated building new network relations. In general, the interviews showed how important previously established personal network contacts are to relocating in exile, although institutionalized networks such as PSI or SAR exist and occasionally can jump in.

Network relations in exile: Home, host, and transnational spheres

Home country

We reviewed the professional network maps from a sectoral perspective. When we distinguished between former home and current host country as well as the transnational realm, we found that the professional relationships with the institutions and their colleagues in their home countries were mainly cut off. A few interviewees retained substantial contacts with colleagues back home, but even then the network ties were stronger in other sectors. Only one interviewee said that she could go back home to her position at the Turkish university because "my university is a very open liberal place" (Interviewee 5). Most, however, had lost contact. Several of the interviewees articulated their disappointment with their home universities. They had been disciplined as a result of their research topics and teaching, thus making the venues with freedom ever smaller even before they had lost their positions.

One interviewee said that "in my university I was told to not use certain words, for example regarding the [Armenian] genocide . . . or do not put this book into my syllabus. . . . And once they had this idea of filming our classes. I freaked out" (Interviewee 9). Often, the perceived cowardice of their home institutions led to immense disappointment and a complete severance of relations (Interviewees 4, 7). One even exclaimed: "The university administration can go to hell" (Interviewee 11). Likewise, the interviewee from Syria said that he cut off relations completely because "my problem in Syria was very political and related to academia" (Interviewee 8). The Yemeni argued that "it is so unstable there that you cannot do anything, so it is meaningless" (Interviewee 13).

Most scholars indicated that, beyond official ties, some contacts had remained in place on a personal level because some colleagues had become friends, but that this did not impact their academic career any longer. The networks that had been left behind back home were referred to as irrelevant by many of the interviewees. Some mentioned that they would still be able to contact their former supervisors and mentors in Turkey, but under the current circumstances, they would not be of much help. Even if joint projects were planned, it would be difficult to work together because of the distance: "You distribute the tasks, and after three months, you meet again via Skype and nothing is done" (Interviewee 7). One scholar explained that she had changed professional relations with people in Turkey. She would rather contract them as local researchers to collect information: "It is a need-based contact" (Interviewee 11). One explanation for this cut in ties can definitely be that they are trying to manage the new circumstances in a different country where the former networks are not seen as so relevant. However, in some cases, there was very clear distancing occurring due to this emotional disappointment.

Host country

In contrast to their home countries, the majority of the scholars indicated that their professional network was more extensive in their host country, Germany. Typically, a lot of significant contacts were project-related. Once a joint activity was planned, such as co-writing an article, planning a seminar together, or submitting a project proposal, the contacts became better established. Thus, they were highlighted as more important on the network maps. Although this might be seen as typical for academic work, some of the scholars suggested that because German academia has a limited number of permanent positions, much of the academic work is short-term and project-related, thus requiring personal contacts. These close contacts were mostly occurring in the exiles' direct environments with the host professors and other senior scholars at the host institutions.

Many of the interviewed scholars had held senior or close to senior positions in their home countries. In Germany, in all cases, the participants mentioned senior mentors and supporters in permanent, leading positions (e.g., directors of institutes or deans of faculties) as their closest and most important connections. It was not clear from the interviews whether all these connections were an academic exchange between equals or considered a hierarchical relationship in which the exiled scholar depended on help from the established German scholar. In any case, most of the interviewees did, indeed, praise their senior hosts for their support. Due to the competitive nature of the German system to obtain project-related funding, it is often important to rely on personal support and guidance from a senior scholar. This strategy had definitely been incorporated into the networking strategies of exiled scholars. One explained that "the German university system is highly closed. . . . So you need to know someone" to get into a project or position (Interviewee 14) and another one said her hosts "bring me into the system . . . They are connecting me" (Interviewee 11). There was only one exception to this dependency: the person from Yemen who argued that "I created my position out of nothing. . . . The people are supportive, but this is not enough. So, I contributed to it" (Interviewee 13). He had brought his international and company contacts with him. He was one of the few participants not researching in humanities and social sciences, but in engineering, which enabled him to also benefit from industrial contacts beyond academia.

For many scholars, institutionalized networks of support, such as PSI or SAR, played a role and were mentioned regularly in the interviews, albeit on the outer circles of their network maps. Some conveyed gratitude toward these initiatives. The Syrian scholar remarked, "SAR created a network for me in Germany" by finding someone interested in his profile (Interviewee 8). Another complained, however, that during PSI workshops, "I feel like a kid, like a primary school student" (Interviewee 12), so she avoided these seminars. One scholar, who seemed to have deeper contacts in these organizations, mentioned that, through a workshop, she found a like-minded scholar in her field with whom she could collaborate (Interviewee 4). She argued that she made a virtue out of necessity. Since her personal struggles had opened up a new research field for her, she was now engaged in promoting academic freedom by lecturing about her situation.

It is important to note that a substantial portion of the strong network ties located in the German sphere were contacts with fellow nationals who were also in exile. Interviewee 9 said that at her Turkish home university: "most of my friends, around 80%, resigned or moved to other places" as she had. The most important colleagues in their field of research and those who shared their political and ethical positions also had to flee into exile. They had now been relocated to Germany or elsewhere; thus, a spatial transfer of relevant network ties had taken place. In particular, the Turkish scholars who were interviewed relied strongly on their BAK network, which now had a major branch in Germany, and they were well connected with most of the colleagues in this network through joint projects such as the Off-University.

One scholar pointed out that professional relations in academia and, in particular, in the social sciences and the arts, need to be based on mutual understanding and even friendship, noting that one cannot work or publish with a person who does not share the same basic convictions. Thus, she called those scholar friends her "allies" and added that "you have to have some confidence in this person" (Interviewee 5). This was also a viewpoint held among the Turkish scholars. The possibility of being able to form a network with other like-minded scholars abroad was specific to the Turkish participants. The two Arab scholars did not mention having these bonds. Typically, related disciplines, a shared political background, and the necessity for immediate solidarity created these strong ties. However, upholding the political aspect of being in exile and being within a specific politically motivated network could also create a burden. Interviewee 11 stated, "I do not want to spend my life with these activities like organizing demonstrations, but they [the Turkish government] push us. . . I am not loving it, but this is something that you have to do." However, another scholar who was involved in some BAK activities in Germany argued, "I actually cannot carry any more problems" and wanted to quit these activities because "psychologically it is heavy for me" (Interviewee 12). Thus, it appeared that such a community-like network can provide strong ties of support but is, simultaneously, a reminder of the fate of exile.

Transnational sphere

The contacts mentioned in a transnational sphere (i.e., someone not located in the home country or Germany) were rather weak and dispersed. Only a few scholars indicated they had an effective transnational network. One interviewee explained that before exile, she had been writing almost exclusively in Turkish; she was only now adapting to the circumstances in Germany by giving lectures and writing in English (Interviewee 3). The section of transnational contacts on her network map was empty. Many other network maps looked similar, usually with only international network organizations, such as SAR, mentioned in the outer circles of the transnational sector. Obviously, for some scholars, these were seen as institutions that provided a gateway to networking, but for only a few did it result in substantial transnational contacts.

However, for three scholars, transnational relations dominated their professional networks or were as important as their contacts in Germany. In all these cases, this was the result of a previous longer stay at a university abroad, such as in the US or Italy. It was often connected to joint work, such as editing a volume or organizing a conference panel together. "I was already working with people abroad and not so much with people in Turkey," explained Interviewee 2, "so it does not feel like exile because there were already so many contacts abroad . . . and I can at least do my research." Importantly, he thought that this situation differed from that of many of the other Turkish scholars.

Some other interviewees considered contacts in the transnational realm as those to "well-known scholars in the field" (Interviewee 10). They had once met them in a particular setting, such as a conference, but still put these contacts at the outer circles of their network map, indicating less direct importance and weak ties. As Interviewee 9 explained, because of their international reputation in a certain field, these scholars were considered key to opening doors in their new Western academic environment, and she had approached the esteemed researcher to write letters of support for her. For her, this helped in getting a scholarship, but it also epitomizes the asymmetric power relations these scholars are in. Similarly, reflecting on this asymmetry, others indicated the desire to approach these distinguished experts for conference participation or a guest lecture spot, or because a particular person "might be helpful to introduce me to other people" (Interviewee 10).

They noted that these people could not be considered strong ties in their networks, but hoped to be able to rely on them in cases of emergency.

Network relations in exile: Communication and activities

During the COVID-19 pandemic, it became common knowledge that spatial distance has a tremendous impact on communication. This research was conducted before the pandemic, but it had already become clear that physical distance strongly affected the intensity and quality of communication. It was also found important by the interviewees whether a contact was considered professional only or a professional-friend relation. Moreover, the restricted communication in their home country was pertinent.

Typically, the more important a contact was based on their position on the network map, the more frequent and the more diverse the modes of communication were. The participants described these contacts as significant because they were perceived as mentors. These people functioned as hosts and/or worked on joint projects with them. Communication occurred in personal meetings (often in person), as well as by means of telephone, messenger, and email. Without the possibility for personal meetings, the ties in the network became weaker, which was observed in relation to the home country but also in the transnational sphere: "Connections to my home country are getting harder, because you need to meet face-to-face. For private contacts it works, but when it comes to work you need to see each other, Skype meetings don't do it" (Interviewee 7). Even those who had a substantial number of contacts in the transnational realm communicated less frequently with them than with their colleagues in Germany. Only in those cases in which a joint project was being carried out were these transnational colleagues contacted more often and via phone (Interviewee 6). Normally, infrequent email was the means of this communication. One scholar mentioned about a colleague in the US that "if I really had to ask something, he would be there" (Interviewee 12), and another said, "these people are so busy" (Interviewee 5), indicating that they were perceived as too important to be contacted frequently.

Many interviewees noted that as soon as personal contacts go beyond a purely professional relation, messenger services were used to stay in contact both in the German and transnational spheres. Whether or not this blurring of private and professional or informal and formal communication was happening depended primarily on the setting between the ego and its alteri.

As Interviewee 13 stated, "If it is more private communication, or an urgent matter, I can WhatsApp, but for professional contacts, it is mainly email," adding that sometimes phone calls could be used, which applied most often to the host professor. Within the BAK network, the use of messenger services was much more common.

In the interviews, many scholars stated that academic freedom, including freedom of speech, was the best aspect of their exile. In particular, those who researched sensitive issues, such as the genocide of the Armenians in Turkey or LGBTIQ issues, mentioned that they enjoyed the freedom in Germany. Despite the difficulties in job security, several scholars said that they felt freer to conduct their research in Germany.

However, juxtaposed with this freer environment in exile were other factors: having family back home who might face repression, having insecure resident status after their scholarships ended, and the slowly closing door of possibility to return one day to their home country. This affected their private and, in some cases, professional communication: "If I communicated with someone back home, I might put him in a bad political situation," the Syrian scholar explained (Interviewee 8). Many of the Turkish interviewees also noted the need to be careful in their conversations with Turkish friends and colleagues, in particular, regarding topics that were politically sensitive: "Turkey is not like China, but it [communication] is watched . . . I would not contact very often people in Turkey, although I do not suspect the Turkish authorities to monitor my WhatsApp. But I would not mention very critical things via WhatsApp" (Interviewee 2). Although almost all the interviewees denied direct self-censorship, even in an academic context, one scholar stated that "I censor my speech according to those who listen" because of a court case still pending. She gave the example of once being at a conference in Germany and seeing a person from the Turkish government taking notes, "So you become more careful" (Interviewee 3).

Conclusions

A QNA of 14 exiled scholars in Germany revealed that only two interviewees – the engineer from Yemen and, perhaps, one interviewee from Turkey who still maintained a position at her Turkish home university – perceived themselves as having the "network-making power" theorized by Castells (2011). All the other interviewees indicated a strong dependency on their host professor.

Consequently, this relationship was perceived as the most important in their network. Although this dependency was probably not intended by the host professor, the differing academic environment of Germany, in which permanent positions are rare and competition for limited resources can be intense (see Vatansever in this volume), did become a main factor that influenced the exiled scholars' networks. With their funding schemes and conditions, the support networks seem to reinforce this dependency and perceived power asymmetry, although their intentions might be different.

Obviously, in an imbalanced global knowledge production system, previous network relations to colleagues and mentors remaining in the home countries often became irrelevant for these scholars, even though digital media facilitated communication. Most networks had to be built anew such that the exiled scholars often felt dependent and had limited capacities to be "network-makers". Help from institutionalized networks in the host country was appreciated and often, well used, but it became clear in the interviews that many were tired of being considered a scholar at risk who had to be "saved." Yet, many of the Turkish scholars succeeded in forming networks in which like-minded exiles in Germany became their most important contacts. Perhaps in the long term, these scientists will obtain network-making power on their own terms. Networking is also a potential source for bringing otherwise unavailable knowledge, ideas, and experiences into the German system, thus enriching it with truly distinct perspectives. However, due to the strong structural dependencies of the German academic system and the inability to exploit its home and transnational contacts in a more substantial way, the networks of these scholars in exile remain vulnerable.

References

Ahrens, P. (2018). Qualitative network analysis: A useful tool for investigating policy networks in transnational settings? *Methodological Innovations*, January-April, 1–9. https://doi.org/10.1177/2059799118769816

Aktas, V., Nilsson, M., & Borell, K. (2018). Social scientists under threat: Resistance and self-censorship in Turkish academia. *British Journal of Educational Studies*, 67(2), 169–86. https://doi.org/10.1080/00071005.2018. 1502872

Bilecen, B., & Amelina, A. (2017). A network approach to migrants' transnational biographies. Working paper. Frankfurt: Goethe-University.

Retrieved from https://www.fb03.uni-frankfurt.de/66520212/Working-p aper-No_12.pdf

Brandstädter, E.-M. (2019, July 8). Türkische Wissenschaftler im Exil: "Kurz vor dem Kollaps" [Turkish scientists in exile: „Close to breaking down"]. *taz - die tageszeitung.* Retrieved from https://taz.de/Tuerkische-Wissensch aftler-im-Exil/!5605367/

Burt, R. S. (1984). Network items and the general social survey. *Social Networks,* 6(4), 293–339. https://doi.org/10.1016/0378-8733(84)90007-8.

Cantwell, B. (2011).Transnational Mobility and International Academic Employment: Gatekeeping in an Academic Competition Arena. *Minerva,* 49(4), 424-55.

Castells, M. (1996). *The Rise of the Network Society.* Malden: Blackwell.

Castells, M. (2011). A network theory of power. *International Journal of Communication,* Special Section: 'Network Theory', 5, 773–87, https://ijoc .org/index.php/ijoc/article/view/1136

Council for At-Risk Academics – CARA (2017). Annual Report 2016-2017. London. Retrieved from https://www.cara.ngo/wp-content/uploads/2017 /09/170912-CARA-AR-Final.pdf

Fahey, J., & Kenway, J. (2010). International academic mobility: problematic and possible paradigms. *Discourse: Studies in the Cultural Politics of Education,* 31(5), 563–575.

Geertz, C. (1973). *The Interpretation of Cultures.* New York: Basic Books.

Hepp, A., Roitsch, C., & Matthias B. (2016). Investigating communication networks contextually. Qualitative network analysis as cross-media research. *MedieKultur,* 60, 87–106.

Hollstein, B. (2016). Qualitative approaches. In J. Scott & P. J. Carrington (Eds.), *The SAGE Handbook of Social Network Analysis.* (pp. 404–16). London: SAGE. http://dx.doi.org/10.4135/9781446294413

Kettler, D. (2011). A paradigm for the study of political exile: The case of intellectuals. In M. Stella, S. Štrbáňová & A. Kostlán (Eds.), *Conference Proceedings: Scholars in Exile and Dictatorships of the 20th Century* (pp. 204–17). Prague: Centre for the History of Sciences and Humanities of the Institute for Contemporary History of the ASCR.

Kreis, Y., Nierobisch, K., & Weber, C. (2019). Netzwerke & akademische Karrieren (Networks & academic careers). In S. M. Weber, I. Truschkat, C. Schröder, L. Peters & A. Herz (Eds.), *Organisation und Netzwerke: Beiträge der Kommission Organisationspädagogik* (pp. 301–310). Wiesbaden: Springer. https://doi.org/10.1007/978-3-658-20372-6_28

Löhr, I. (2014). Solidarity and the academic community: The support networks for refugee scholars in the 1930s. *Journal of Modern European History*, 12(2), 231–246. https://doi.org/10.17104/1611-8944_2014_2_231

Scheibelhofer, E. (2006). Migration, Mobilität und Beziehung im Raum: Egozentrierte Netzwerkzeichnungen als Erhebungsmethode. In B. Hollstein & F. Strauß (Eds.), *Qualitative Netzwerkanalyse: Konzepte, Methoden, Anwendungen*. (pp. 311–331). Wiesbaden: VS.

Vertovec, S. (2002, 14-15 February). Transnational networks and skilled labour migration. Conference Ladenburger Diskurs "Migration" Gottlieb Daimler- und Karl Benz-Stiftung, Ladenburg.

Vertovec, S. (2010). Transnationalism and identity. *Journal of Ethnic and Migration Studies*, 27(4), 573–82. https://doi.org/10.1080/13691830120090386

Towards structural responses to the displacement of scholars
The Mapping Funds project

Asli Telli

> Academic freedom is individual, collective and institutional... In fact, academic freedom is intellectual freedom in academic roles and contexts (Moshman, 2017, p. 12).[1]

War and anti-democratic pressure impose unprecedented impacts on freedom in academia, once again reminding us that academic freedom is an indispensable part of human rights. Since 2015, first the migration crises because of the war in Syria, then the coup attempt in Turkey and its discriminative domestic politics, and finally with the aggregation of political pressure all over the world, this issue has been reignited in Europe. In many parts of the world, scholars have lost their jobs, been imprisoned, and/or have had to flee, while others are still working in their home countries under authoritarian rule, facing life-threatening risks. One example is the political oppression of the Academics for Peace (AfP) organization from Turkey. The AfP network comprises around 2,200 academics who have signed the petition entitled "We will not be a party to this crime" (Academics for Peace, 2016)

---

[1] Moshman refers to an interdependent and multi-layered definition of academic freedom in this statement. He further clarifies: "Academic freedom exists at multiple levels. Its legitimacy at each level depends on whether it protects academic freedom at all levels. The academic freedom of individual faculty members should be respected provided they are respecting the academic freedom of their students. The academic freedom of the collective faculty should be respected provided they are respecting the academic freedom of individual faculty members and students. The academic freedom of academic institutions should be respected provided the institution is respecting the academic freedom of its faculty and students, and thus operating with academic integrity."

and who have been relentlessly criminalized by the Turkish government as traitors, terrorists, and complicit supporters of the Kurdish struggle for freedom at the southeast borders of the country.

Since the launch of the petition in 2016, many organizations, foundations, federal and state bodies, universities, and institutions in Europe have declared their support of AfP.[2] It is noteworthy that the case of AfP is only one of many rights violations and interventions against academic freedoms in the last decade. Thus, the structural responses need clear analysis without geographical or contextual limitations.

The chapter examines the third-party funding sphere for at-risk scholars. Third-party funding, such as postdoctoral and other scholarships, has grown considerably since 2015, and different initiatives have been created to support at-risk and refugee scholars. My analysis is based on the Mapping Funds project, which was created in October 2017 and assesses different funding and support instruments. We then focus on similar projects/initiatives and eventually arrive at the identification of the structural gaps that exist in the landscape of current support and self-help initiatives for these scholars. Finally, the chapter demonstrates possible trajectories for overcoming these gaps through bottom-up, collaborative, digital platforms of action that facilitate peer dialogue and transnational solidarity.

I will begin by presenting an analysis of the Mapping Funds project,[3] which I supervised and coordinated from 2017 to 2019, to illustrate the state of the art in the support and funding ecosystem for displaced researchers. This mapping effort also introduces current issues and inquiries into the third-party funding landscape. In a general sense, there are a number of ways to provide support to at-risk and displaced scholars: Providing funding and refuge for research, volunteering for proofreading initiatives, supporting campaigns for persecuted researchers, and joining advocacy campaigns for

---

2   One example of such support came from the Academia for Equality members' organization of 400 academics dedicated to advancing the equality and democratization of Israeli academia and society: https://afp.hypotheses.org/372. Another example is from BdWi, The Association of Democratic Scientists in Germany, which sent a support letter addressed to Heiko Maas, the Minister of Foreign Affairs at the time, to bring attention to the repression against critical scientists in the AfP network: https://www.bdwi.de/show/10707775.html

3   The Mapping Funds project (https://www.mappingfunds.com) was generously funded by the Consulate of Sweden in Istanbul and supported by auxiliary research funds from the Alexander von Humboldt Foundation PSI program from 2017–2019.

global academic freedom are the most noteworthy. More detailed support options are accessible on the Scholars at Risk (SAR) website.[4] Often, initiatives such as SAR call for "network diversity," that is, different forms of support coming from various institutions such as federal governments, foundations, universities, and peer networks. This so-called network diversity has its advantages, but as the results of the Mapping Funds project illustrate, this diversity can also have its drawbacks. We will also identify structural gaps in the support landscape by referrals to the current findings in the SAR's *Freedom to Think 2020* report, the Global Public Policy Institute's (GPPi) Academic Freedom Index 2020 report (Kinzelbach et al., 2021), and the InspirEurope 2020 report, entitled *Researchers at Risk: Mapping Europe's Response* (EUA, 2020).

Mapping Funds as a case study

The aim of the Mapping Funds project, which was kickstarted in October 2017, is to analyze the support network and funds available to at-risk and displaced scholars in Europe. The project exists for the purpose of generating a road map for scholars and to discuss the institutional and grassroots roles for enhancing open spaces for knowledge exchange. The initial focus has been analyzing the networks of support and funding received by at-risk scholars from Turkey since 2016. However, during data collection, researchers' encounters with funding institutions have shown that the support and funding networks cannot be limited to a specific geography and need to cover the mobility of displaced scholars from other locations to Europe, taking into account the tensions of the war, poverty, and climate change as reasons for the flight and displacement of scholars.

To expand the project's geographical span, the research team chose the *2020 Scholars at Risk Report* as the best tool. The approach in that report within the framework of structural responses was developed in the form of institutional recommendations to geographies where pressures and violations against academic freedom are at higher levels. As a concrete case in point, in the *Global* section of the report (Scholars at Risk, 2020, p. 114), SAR urged states, higher education (HE) leaders, civil society, and the public

---

4   For individuals who would like to get involved, go to: https://www.scholarsatrisk.org/ get-involved-individuals/; for organizations interested in supporting, go to: https://ww w.scholarsatrisk.org/get-involved-institutions/

at large in all countries to "ensure the security and integrity of virtual higher education spaces, assist threatened scholars and students and contribute to efforts aimed at reinforcing principles of academic freedom and institutional autonomy," among other recommendations. In the case of Turkey, SAR urged the state authorities and higher education leaders to "suspend and reverse actions taken against HE institutions and staff, ensure an effective and transparent review and due process for all HE staff and restore and strengthen institutional autonomy." Thus, the global call was made to the public and the states at large, while the call to Turkey was mainly directed to the state of Turkey concerning its complicit measures with the authoritarian government.

Keeping this significant paradigm in mind, the research team expanded their map to include funds and support networks available in Europe in connection with transatlantic ties so that the organizational network would be holistic and would be useful for at-risk scholars in general. The team also especially kept their focus on funds available to at-risk scholars from Turkey since the case of Turkey had unique characteristics in terms of the rigorous assault on academia. This way, the project could then serve the internal needs of the AfP network without excluding the larger cohort of at-risk scholars in other geographic areas seeking support in Europe as well as outside Europe with transnational ties to European institutions. Under these conditions, two different maps have been created on the Graph Commons network mapping platform to illustrate support mechanisms: one of them publicly available, the other one private due to the confidentiality of data involving at-risk scholars.[5] The public map, entitled "Support Networks of At-Risk Scholars," illustrates the general overview of declared funds and support mechanisms by the institutions; it is publicly accessible and open to public contribution. The private map, entitled "Mapping Funds for Endangered Researchers from Turkey," shows the funds and their institutional supporter network, attributed specifically to at-risk scholars from Turkey; it is only available to the project team due to data protection and confidentiality measures. This second map is regularly updated with the collaboration of supporter institutions, either filling in the form on the website[6] or sending the research team email updates.

---

5   The Graph Commons platform can be accessed at https://graphcommons.com/
6   The form on open calls and possible contributions can be accessed at http://mappingf unds.com/en-US/Collaborate

The purpose of the network map is to graphically illustrate the relationality among different actors. In this realm, actors are taken to mean different funding and grant-giving organizations. After conceiving the relationality, it is possible, in a network map, to recognize leading/core actors and understand the central nodes as well as the intermediary pathways, bridges, and actors at the edge with periphery roles. The *network visualization approach* facilitates the investigation of key points in the subject matter and the actual research question. The first step of the project was collecting the data from websites via email inquiries to funding organizations and from personal connections. The public map contains only public data from websites. The private map includes both publicly available data and information retrieved via email contacts with various institutions. After data collection, the *Graph Commons* online platform was used for visualization. This platform was chosen since the software it runs on is open source, and it abides by the tenets of scientific integrity and ethical guidelines of the digital network mapping method; it is also familiar among those in the scientific commons in Turkey.

During the visualization, the idea was to clearly understand the main actors and their relations. To be well prepared for later stages of the mapping framework, specific details such as gender, level of academic career, and research area focus were also embedded in the map as "properties," a platform feature used for filtering and deeper investigation. The last part of the project covered an overall analysis of the maps to frame the research and, more specifically, to define the structural responses and underlying gaps. A structural response refers to fellowships available at host institutions, that is, universities or research centers, and the funding offered by each organization. A few examples of underlying gaps are related to the accessibility of these funds, challenges regarding matching of the at-risk scholar with a host, and the limited duration of stipends.

On a techno-methodological level, the web visualization approach on its own has its challenges in terms of its legibility, interpretability, and accessibility of the host interface. As a result, the following narratives are the priority for Mapping Funds:

- Communicating implicit but feasible funding structures to donors, supporting organizations, funders, and scholars,
- Identifying matching and bridging complementary support mechanisms (e.g., infographics and kits),

- Dissemination of vital information for social good in altered media formats (audio-books for the information and communications technology (ICT)-poor, multimedia trailers for ICT advanced areas, and hybrid publications),
- Revitalization of exchanges between the Global North and South by using scalable maps for highlighting network relations and defining gaps among diverse support mechanisms,
- Simple, legible design for discussion and exchange, and
- Possibility to crowdsource for sustainability through a protocol for the Graph Commons and Mapping Funds interface. Such a protocol would enable interoperability and provide a dashboard for visitors to the Mapping Funds webpage so they can update the site with information regarding new calls and opportunities.

These priorities have been met to various degrees, except for the last item on crowdsourcing. Even though the host interface is Graph Commons with a relatively simple template for network mapping, the project team has not been able to create a community of interested potential mappers. Thus, prior to the final reporting in 2019, the team considered options for embedding customized devices preceded by the protocol mentioned above, or possibly curating a customized app for potential mappers among knowledge workers and third-party funding staff. However, this was not possible due to lack of funding and other life priorities of the core research team. It is important to stress that the team remains open to ideas and potential funders for sustainability.

Figure 1 provides a snapshot of the Mapping Funds digital interface as a working space and a hub for collaboration.

Figure 1. Interactive platforms – working space for confidential collaborators and institutional agents. The Mapping Funds website backend. The interface and backend of the Mapping Funds website provides a customized repository and tools for individual as well as institutional collaboration.

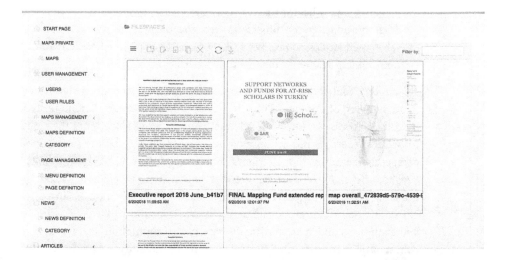

The findings and outcomes of Mapping Funds will be further discussed in terms of the other related initiatives, programs, and projects in the parts to follow. The funds and support networks' landscape has been covered in a relational network in the project to gain a view of the structural responses within the existing power structure and to identify the underlying gaps.

Structural responses within international trajectories

While displaced lives of at-risk scholars involve both short- and long-term implications, they are also a part of international debates and studies around academic freedom. In that regard, the Academic Freedom Index (AFI) (Kinzelbach et al., 2021) provides a state-of-the-art mapping of academic freedoms around the world, crowdsourced by collective efforts of scholars nominated as experts for specific regions. I consider the index a roadmap for what can be done by a plurality of actors in the higher education and human rights fields to ameliorate the dignity of scholars at large. Here is a conceptual description taken from the introductory text of the 2020 report:

> The Academic Freedom Index (AFI) provides manifold opportunities for research, but also for policy debates among government officials, parliamentarians, research funders, university administrators, academics, students, and advocates alike. This report aims to inform such debates. After introducing the objectives and the dataset, it provides specific recommendations on how key stakeholders can apply the index to protect and promote academic freedom (Kinzelbach et al., 2021, p.4).

This report's contribution to operationalizing academic freedom is significant for a number of reasons. First, it has to be made clear that in related legal frameworks worldwide, academic freedom must also secure the right to political exercise and action as long as such exercise points to freedom of speech regarding human rights, public good, and civic responsibility. Second, the right to political exercise and action, when undertaken by academics, must not jeopardize their status, job, or position in any way. Third, this political exercise and action must be for the public good without any exceptions and must not incite hate or discrimination against certain communities. Fourth, academics' freedom for mobility to perform their research and teaching must also be guaranteed by transnational as well as national laws. Mapping the state of academic freedom worldwide is, however, a complex endeavor–due to not only diverging definitions but also the difficulty of assessing whether it provides a sufficient base to debate specific support mechanisms for displaced scholars. Support mechanisms and funding attributed to the research and teaching of academics in the higher education sphere must take the principles of academic freedom as a basis, integrate them into their infrastructure, and provide necessary structures for safeguarding decent research, teaching, learning, and public outreach.

In the Mapping Funds project, we show that the abovementioned support mechanisms form organic networks and contingencies with other institutional actors on the edge (nodes as central actors and edges as actors at the margins). This creates autonomous spaces for collaboration and academic exchange. However, there is a significant catch at this point. The governance of higher education in the Global North has been designed with a hierarchical mindset in order to maintain its rigorous power structures, that is, higher education institutions (HEIs) of the twentieth century and the modern state

(Zapp et al., 2018, pp. 7–8). In that regard, the autonomy of the university is in perpetual contention that exists among the managerial powers, the state, and the third-party funders. The last two decades have introduced new aspects to this scheme with the advent of internationalization in higher education. However, the authoritarian purge of scholars, enforced displacements, and exile, mostly concentrated in the Global South, have caused asymmetric challenges. Thus, the structural responses as the main theme in this chapter refer to the network relations of support mechanisms in today's higher education sphere attributed to displaced scholars, the power dynamics caused by these relationalities, and their impact on displaced scholars. Further to be explored are international trajectories facilitated by the friction among conventional, democratic HEIs, third-party funders, some of whom have neoliberal stakes, and scholars seeking autonomy.

In this regard, collaborative institutional schemes are significant. Such schemes become noteworthy when the main supporting actors collaborate with the dispersed edge actors in the network to form alliances of support. An example of that would be the collaboration between Off-University[7] as an edge actor and host institutions as core actors in Germany. Through tandem teaching in host institutions, displaced scholars continue their teaching careers while they co-teach remotely with a colleague in their field. The funding for this initiative comes from a few prominent organizations in Germany, most recently the New University in Exile Consortium and the Alexander von Humboldt Foundation via the Philipp Schwartz Initiative (PSI) lump-sum budgets of research fellows. The following figure, taken from the Mapping Funds project's extended report, dated 2018, depicts this situation well.

---

7   Off-University creates new strategies to uphold and sustain academic life and knowledge threatened by anti-democratic and authoritarian regimes. It was established for and by academics from Turkey yet addresses itself to academics all over the world: academics who have been purged from their institutions, forced to resign, who are legally and politically persecuted and even imprisoned because of their opinion and research. For further info: https://off-university.com/en-US/page/about-us

Figure 2. *Reading the Map: Institutional Network Collaboration featured with infographics. Mapping Funds 2018 Report. The intersecting parts of these circles refer to collaborative schemes among national and federal state actors, municipal actors, civil society organizations, and charities' support (labeled as autonomous) and the higher education landscape. The mentoring schemes provided by host institutions would be a good example of such collaboration since their funding comes from multiple actors in clusters represented by colored circles. As a concrete example, the Alexander von Humboldt Foundation (AvH) and Deutsche Forschungsgemeinschaft (DFG; German Research Foundation) are the most relational actors, and they both have bridging roles among municipal support, PSI clusters, and state support. The Network of European Institutes for Advanced Study (NETIAS) is another important bridge actor at the edge as it connects the clusters of the universities, Université Catholique de Louvain, and third-party funding. It also plays a significant role as an EU organization to connect actors from other countries.*

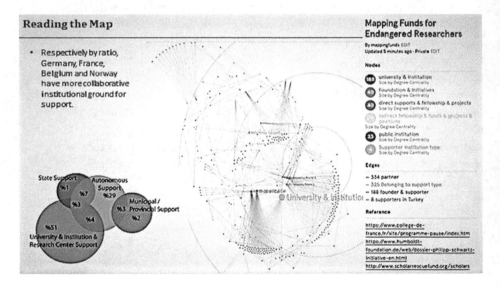

The Mapping Funds project outcomes related to support networks and funding mechanisms would be helpful in expanding on structural responses. Collaborative schemes such as the mentoring schemes by host institutions or peer networks like Academics in Solidarity (AiS) are significant because the varying needs of displaced scholars can be met in time. Related program staff must keep in mind that the displacement of these scholars lasts much longer than the initially planned one-to-three-year stipends and consider options more durable than short-term bridging funds (Telli-Aydemir & Diner, 2021). This places the responsibility on universities and research centers as host institutions, hosting mentors, and institutions for the funding and support landscape.

As for the responsibility of institutional actors, the Global North–South nexus is an imminent framework that requires dialogue among diverse actors; thus, the project team organized two face-to-face workshops (one in Brussels with EU commissioners and the other in Kassel with the Off-University core team and volunteers) to elaborate the issues at stake and continued the dialogue with larger networks of workshop participants through online interactions. Both project workshops as well as the correspondence with staff in funding institutions revealed that at-risk scholars face challenges regarding integration into the HE landscape in the countries in which they settle. The main causes of these challenges are lack of fluency in the primary academic language, lack of academic networks, low literacy in systemic relationalities, and lack of social relations in their daily lives. In the case of the displacement of scholars with families, these challenges cause distress for all family members and an extra burden on the displaced academics. In the case of individual displacement, loneliness and isolation factors may trigger mental health problems and even lead to trauma.

However, in addition to outlining the challenges faced by displaced scholars, the Mapping Funds project also identified the benefits of successful integration of displaced scholars in the host country and the benefits for the development of its research system. Among them are the revitalization of exile knowledge as a significant game-changer in the path toward decolonizing higher education.

The following are the issues arising in interactive platforms initiated by autonomous exile networks:

a) Well-being at the level of human dignity (alliances with civil society, support networks, and HEIs to prevent human rights violations and the shrinking of critical thought as well as the emancipation of at-risk scholars) (Civil Rights Defenders, 2018),

b) New perspectives for internationalization (DAAD, 2020) (formulating policies that integrate at-risk scholars into the international HE landscape),

c) Transitional disciplines across borders (better science communication and vital dialogue in cross-cultural intersections) (Philipp Schwartz Forum, 2020), and

d) Dealing with trauma (daily life challenges as well as resilience of scholars and their families, socio-psychological sharing and caring initiatives, memory workshops, transversal stories, and narratives that create empathy) (Telli, 2021).

These issues are also emphasized in the GPPi report on the AFI, in which they are categorized as effective practices for cooperation among organizations in the academic freedom network. It has been stated that organizations in democracies must develop more mindful, thoughtful, and well-informed cooperation procedures to address the risks of cooperation in an increasingly volatile environment and still maintain productive cooperation projects (Kinzelbach et al., 2021, pp. 22–23). According to GPPi, doing so requires a clear vision of the objectives of cooperation as well as the risk assessments and other due diligence measures that need to be in place before cooperation projects can launch. In line with this principle, and resonating with internationalization, gender diversity/inclusion/equality perspectives of European universities, training for career development, and mentoring networks such as Frauenspezifisches Mentoring Siegen (FraMES)[8] or those focused on at-risk scholars (Career Mentoring-SAR at Uni-Cologne) have been established. Thus, one can say that the debate on academic freedom worldwide and the improvement of support mechanisms for displaced scholars are strongly complementary and have to be discussed in the same context.

It is necessary to discuss the role of specific institutions. The responsibility is not only on funding institutions alone but also on host institutions and mentors. According to Mapping Funds mapping and monitoring reports, welcome offices in most hosting institutions have created special opportunities for individual and family counseling. However, peer networks are a significant missing factor in this picture. While it is true that such peer work must be at a grassroots level and initiated by the at-risk scholar community itself, the structures of hosting institutions, such as common spaces for socialization and exchange, assisting staff, and compensation

---

8   FraMeS – Frauenspezifisches Mentoring Siegen: https://www.uni-siegen.de/gleichstell ung/karriere/frames/index.html

budgets, are definitely not there. With many challenges to overcome upon settlement in a new country, at-risk scholars need a long time before they can make contact with their peers and close communities for the existing capacities of the country, the means to collaborate, and the available resources to support their well-being and resilience.

In this regard, it is worth examining initiatives that have been created elsewhere. A case in point is the International Society of Political Psychology (ISPP) Twinning Program as a special effort to help scholars from Turkey working in the field of political psychology (ISPP, 2017). ISPP members among scholars under threat are matched with a volunteer mentor, also in the member pool, as a way to show solidarity as a partner. It is an attractive opportunity for political psychologists since they get the opportunity to build transnational alliances and collaborative knowledge production when facing difficult times during their career. The distinct characteristic of this effort is the three-tier approach: 1. The program offers support for "letter writing" to authorities when a member scholar faces threat; 2. A donation scheme provides small funds for career support; and 3. A job matching scheme helps to find potential hosts for scholars under threat. The scale of this program is limited; however, it provides a working model and inspiration for potential peer networks in other disciplines.

As another case in point, Academics in Solidarity (AiS) also features a solidarity network, but it is positioned more in intra-university cooperation compared with that of the ISPP. It is administered at Freie Universität Berlin, and the following is stated on their own website:

> AiS is a peer-mentoring program that connects exiled researchers and established scholars in Germany, Lebanon and Jordan. It seeks to create a network of solidarity, strengthen the value of cross-cultural research cooperation and open up new perspectives within the academic environment of the host country.

The main services provided by AiS are mentoring, academic counseling, systematic network building, and funding of small research endeavors. Thus, with its trans-regional outreach for creating safe zones for knowledge exchange and career support for endangered scholars,[9] AiS touches upon

---

9   The Knowledge Hub of AiS is an active repository of useful information for at-risk researchers: https://www.fu-berlin.de/en/sites/academicsinsolidarity/Knowledge-Hub/index.html

significant challenges of academic freedom and mobility. The recent efforts of the network for making policy recommendations to relevant higher education actors and their encouragement of active engagement of at-risk scholars show that the issues at stake require durable solutions and active dialogue on the part of all intramural and extramural actors involved.

Exclusive responses, gaps, and challenges

As reflected in the InspirEurope report as well as the Mapping Funds executive summary, the EU does not currently offer a dedicated fellowship program to at-risk scholars. A further complication is that while at-risk scholars technically have access to EU funding schemes for international researchers, there is clearly a lack of awareness of these possibilities. It is true that current efforts, such as the InspirEurope project, are addressing this issue of awareness; however, it is not completely up to support organizations and (potential) host institutions per se. The findings of the InspirEurope project suggest that the eligibility criteria and the high competitiveness cause low application and success rates among at-risk scholars. It is further emphasized that the requirements and conditions of European programs might hinder or delay at-risk scholars' integration and career development. The best example of this is the mobility restriction under the Marie Sklodowska Curie Actions of the European Commission, wherein the researcher may not continue residing in the host country of the first settlement while carrying out the project. Especially once researchers unite with their families, resettlement to a new country becomes a more complex challenge.

As for the self-reflective findings in the InspirEurope 2020 report, the top five challenges identified by researchers at risk in Europe all relate to employment and professional development. Here is a summary of these challenges rated as "very challenging or challenging" according to the report:

- Finding employment that matches their academic and professional background (84%),
- Finding employment in general (77%),
- Temporary nature of placements and employment contacts (75%),
- Issues around the resulting transitions into different employment (65%), and
- Learning the local language (64%).

The main challenge noted was making a smooth transition to academic careers or other positions in the host country. The issues surrounding this challenge are also striking. Especially when no prior foreign language skills exist, learning the local language up to a level favored by the job market requires a long time and may be a burden along with other duties. The cultural drawbacks may also be a determinant in the social isolation of newcomer researchers and prevent them from feeling a part of the surrounding community.

In terms of policy recommendations, the InspirEurope project stands out as an initiative that supports, promotes, and integrates researchers at risk, which consists of 10 partners. The project's recent report provides a comprehensive framework. In a survey distributed to at-risk scholars and supporting institutions, which received 260 responses, there was a special category of "portals, services, support mechanisms used by researchers at risk." Under this category, the services available were identified as specific services for at-risk researchers/refugee scholars/asylum seekers, general portals for researchers/immigrants, national/local support services, and informal networks of peers and support via civil society organizations. These services were evaluated by researchers via an online survey, and the results were reflected in the report as such:

> When asked to rate the portals, services, and support mechanisms they used, those targeting specifically at-risk researchers were rated as (very) good, and 21% said these could be improved. As for general portals, a further 41% gave positive feedback on portals or services specifically addressing those in a refugee-like situation; however, around one-third were not aware of any such services or portals. Informal exchanges, such as peers and informal networks, were highlighted as one of the most useful and frequently used resources and play a crucial role for many of the respondents. 80% of researchers have in the past gathered information and support via an informal network of peers, but 44% said that these could provide even better support if enhanced (14% not sufficient, 30% could be improved), perhaps with a more formalized network structure (EUA, 2020, p. 32).

As mentioned earlier, from this brief finding, one can easily infer that supporting peer and informal networks is essential. In the same report, these portals, services, and support mechanisms were categorized as those provided by hosts and those by support organizations and projects. While those provided by the hosts aim at career/capacity and skills development, such as language and teacher training or job placements, support organizations focus more on fellowships, networking, matchmaking, and mentoring schemes.

Resetting lives and settling into new socio-cultural environments may create unexpected externalities for displaced researchers. Additionally, due to short-term funding opportunities, there is a substantial need for complementary support mechanisms. Thus, cooperation among different institutional actors in the network is required to operationalize well-defined support with intermediary actors, as in the case of Scholars at Risk NY and Europe. For example, the SAR main office in New York works with other SAR liaisons in Europe, the universities in Europe, the federal state agencies, and career development organizations in order to help sustain the academic careers of at-risk scholars.

A few of the mechanisms that are used to tackle externalities are explained here to give an idea about what can be done in a decentralized fashion. One such example is AuthorAID,[10] a free pioneering global network that provides support, mentoring, resources, and training for researchers in low- and middle-income countries. It supports over 20,000 researchers in these countries by helping them to publish and communicate their work. The network offers personal mentoring by highly published researchers and professional editors, online training workshops on scientific writing, a forum for discussion and questions where researchers can benefit from advice and insight of members across the globe, access to a range of resources on writing and publishing, and networking. AuthorAID also has a southeast Mediterranean branch that offers publication support and editing services.

Another interesting case in point is a peer network called Chance for Science,[11] which was initiated by Carmen Bachmann, a professor at Leipzig University. Even though there was no interest from refugee scientists in the initial stages, as a result of Bachmann's personal efforts, there are currently more than 500 subscribers, of which one-quarter are refugee scientists. The related search and profile match platform was realized by Bachmann's graduate students. She is personally engaged in the endeavor as she mentions in the podcast; she paid the first visit to a formerly abandoned building (former psychiatric hospital) where 1,000+ refugees lived during

---

10   For the website: www.authoraid.com; for the East Mediterranean branch: http://www.authoraidem.org/

11   For the website: https://home.uni-leipzig.de/~chanceforscience/index php?lang=en

their application for residential status. She called to see whether there were any scientists/researchers in the house and handed the self-made flyers to some of those who approached her. Still, the platform she initiated for exchange between refugee researchers and their German colleagues was not frequented until almost a year had passed; nevertheless she continued her efforts to raise awareness about its significance. It was as late as 2019 that the platform became known among refugee scholars and their German colleagues interested in research collaboration. Thus, this recent experience[12] is an indicator of how much daily life's struggles can get in the way of keeping up with research and academic careers in the case of displaced researchers. It also underlines the need for awareness-raising both among the newcomers as well as the German academics eager to show support and solidarity over the long term.

According to the Care Advancement for Refugee Researchers in Europe (CARe, 2020) project findings, continued existential uncertainties due to prior risks as well as ongoing precarity often have an impact on the mental health and well-being of researchers at risk and their families. The psycho-social counseling units at universities provide good opportunities for regular mental health care for the researcher and their family at least in the first years; however, a sustainable solution would be better.

Toplumsal Dayanışma İçin Psikologlar Derneği (TODAP; Psychologists for Social Solidarity) in Turkey is a good case in point. The TODAP community offers pro bono counseling and open seminars for communities in need; they have also shown solidarity with AfP since 2017. TODAP volunteers also offer individual counseling. Another recent example from Turkey is the Birarada Association, which launched its digital platform in June 2020, thanks to its supporting organizations. Birarada unites all solidarity academies in Turkey, initiated by dismissed academics in the AfP network. The digital platform also hosts the Mentoring Support Working Group, which currently has around 300 members. The group recently organized well-being seminars comprising three different sessions, which were attended by interested parties. The pandemic has apparently created an extra mental burden, so the extended lockdown periods can be a good time to connect with peers. The first event of the working group was an experimental one to determine what comes next

---

12   For a more detailed account of the story and the podcast: https://home.uni-leipzig
     .de/~chanceforscience/index php?option=com_content&view=category&layout=blog
     &id=21&Itemid=160&lang=en

and what the communities' needs are. Working group members are eager to invite other knowledge workers – authors, poets, and artists from different geographies to the well-being events. Especially for graduate students, the lockdown situation in tiny on/off campus student flats has been very tough since March 2020. Thus, peer support groups are essential for graduate students and a high priority for postdocs and senior researchers.

According to the findings of Mapping Funds as well as other cited and mapped projects in this chapter, bottom-up action toward policy formulations is essential for creating and sustaining decent work and life conditions for displaced scholars. Thus, a new initiative called Share the Platform[13] founded by a group of scholars and practitioners with refugee and non-refugee backgrounds, calls upon refugee colleagues to deliver their expertise on the unique needs and experiences of refugee populations in a wide variety of fields. Founded in the US and endorsed by SAR, they aim to consolidate the expertise of refugees to improve policy, programs, and practice. The platform is a transnational advocacy hub and acts as a model that serves as a point of exchange and training for institutions and people from refugee backgrounds to move toward full partnerships to create meaningful change in program design, policymaking, and action. Such consolidation efforts for expertise and experience of displaced scholars and practitioners in policy-related fields must be multiplied. The collaborative work of scholars with practitioners in the field would also be an asset in the efforts toward achieving better policies, thus illustrating how the expertise of scholars with refugee backgrounds animates the entire higher education ecology.

Concluding remarks

This chapter focused on structural transnational responses to dismissed and repressed academics whose lives are displaced in some cases geographically and in some mentally, while in other neglected cases, in both ways. The prominent issue regarding this focus is "finding a way out," both for career sustenance as well as for social lives. As highlighted by Vatansever (2020, pp. 127–128), the general tendency is to create fragmented and widespread micro-organizations to challenge a huge and multi-centered systemic power. Social movements of our time, as well as the sphere of intellectual subjectivity

---

13   See https://www.sharetheplatform.org/

and resistance from within the ranks of academia, have adopted this formula on an ad hoc basis. Vatansever defines this formula of resistance as one that creates networks of solidarity based on shared precarity and further specifies that most of her interviewees agree with this definition. While this resistance provides endurance and hope, one must not forget that solidarity academies, critical knowledge networks, volunteer mentoring schemes, and mutual support groups are faced with major systemic challenges. After presenting the results of the Mapping Funds project and a presentation of additional bottom-up experiences to support displaced scholars, I believe there is a need for sustainable action plans that include the following support mechanisms:

- Remote fellowships, including digital repository and library access[14] where mobility is restricted, keeping in mind that brain drain of knowledge workers may create a detrimental impact in authoritarian states where all freedoms are under attack.
- Grants, funds, and special programs for graduate students who are facing similar threats in their early careers must be increased. The recent DAAD scholarship initiative (Hilde Domin Programme) (Academic Cooperation Association, 2021) is a good example in this regard.
- Considering the fact that decent labor conditions and diversity are common problems in the current higher education landscape, critical space-place pedagogies, which facilitate the coalescence of social justice endeavors, must be curated. This would make room for more meaningful alliances among displaced scholars, their colleagues, and peer networks.
- Increasing lecturer/teaching positions and research opportunities for specific area studies in a Global South–North comparative perspective would be a feasible plan in this regard.
- Host institutions must take responsibility for the career sustenance and well-being of at-risk scholars as well as their families. Keeping in mind the fact that fellowships and stipendium programs provided by foundations and federal state structures are limited in time and scope, the academic community, as well as the services in universities and research centers,

---

14   A case in point in this regard is the New University in Exile Consortium efforts. For more: https://newuniversityinexileconsortium.org/programs/programs/ and https://ne wuniversityinexileconsortium.org/programs/library-access-initiative/

must be prepared for the long-term integration of these scholars (Telli-Aydemir & Diner, 2021).

- As mentioned in this chapter, significant steps toward relevant policies are already in place; however, the involvement of at-risk scholars themselves in the policymaking process is a must. Only then can the actual needs and mutual expectations be identified and related measures be devised in terms of democratic equity.

Thus, the outcomes of the Mapping Funds project once again underline that durable policies must be in place to win back human dignities and mitigate the deprivation of displaced scholars. However, the policymaking process must not take place behind closed doors or be controlled by privileged actors. Inclusion of displaced scholars in all stages of policy negotiations and the policymaking process is vital. Thorough mechanisms for active inclusion and open spaces for dialogue are the primary factors that must be considered. On November 29, 2018, the European Parliament, after engaging in negotiations with the AfP international affairs working group, released a special report on academic freedom which stated that human rights must be a consideration in the EU's foreign policy framework. The report calls for EU-funded action to ensure the protection of at-risk students and scholars (European Parliament, 2019; see also SAR, 2018). This report is fully in line with the findings of the Mapping Funds project: It finally opens the way for collective action in Europe toward protecting critical scientific knowledge and recognizing scholars who are the producers of that knowledge as essential and students as its standard-bearers for an informed, democratic future.

References

Academic Cooperation Association (2021). DAAD and OeAD: support for students and researchers at risk. Retrieved from https://aca-secretariat.be/newsletter/daad-and-oead-support-for-students-and-researchers-at-risk/?titleId=1&articleId=4&edition=2021

Academics for Peace (2016, January 10). We will not be a party to this crime! Retrieved from https://barisicinakademisyenler.net/node/63

Care Advancement for Refugee Researchers in Europe – CARe (2020). Labor Market Conditions for researchers with a refugee background: Main findings from selected European countries. Retrieved from https://6kyw

p25ru3q2da9i037dyvc8-wpengine.netdna-ssl.com/wp-content/uploads/2
020/05/Labour_market_conditions_for_researchers_with_a_refugee_bac
kground.pdf

Civil Rights Defenders (2018). Contribution of representative from Civil
Rights Defenders at Mapping Funds outreach workshop. 20 June.
Brussels: Mundo-B.

Deutscher   Akademischer   Austauschdienst   –   DAAD   (2020).
Internationalization in Higher Education for Society: Concept, Current
Research and Examples of Good Practice. Retrieved from https://static.d
aad.de/media/daad_de/pdfs_nicht_barrierefrei/der-daad/analysen-studi
en/daad_s15_studien_ihes_web.pdf

European Parliament (2019). Defence of academic freedom in the EU's
external action. Retrieved from https://www.europarl.europa.eu/doceo/
document/TA-8-2018-0483_EN.pdf?redirect

European University Association – EUA (2020). Researchers at Risk. Mapping
Europe's Response. InspirEurope. Retrieved from https://eua.eu/downlo
ads/publications/inspireurope%20report%20researchers%20at%20risk%
20-%20mapping%20europes%20response%20final%20web.pdf

International Society of Political Psychology – ISPP (2017). ISPP Efforts to
Help Turkish Scholars. Retrieved from https://ispp.org/ispp-blog/ispp-ef
forts-to-help-turkish-scholars/

Kinzelbach, K., Saliba, I., Spannagel, J., & Quinn, R. (2021). *Free Universities:
Putting the Academic Freedom Index Into Action.* Full report. Global Public
Policy Institute – GPPi. Retrieved from https://www.gppi.net/2020/03/2
6/free-universities

Moshman, D. (2017). Academic Freedom as the Freedom to do Academic
Work. *Journal of Academic Freedom, 8.* https://www.aaup.org/sites/default
/files/Moshman.pdf.

Philipp Schwartz Forum (2020). Philipp Schwartz Forum workshops related to
better science communication and diversity in science. 9–10 March 2020.
Berlin.

Scholars at Risk – SAR (2018, November 30). European Parliament adopts
report that resolves to prioritize academic freedom in EU external affairs.
Retrieved from https://www.scholarsatrisk.org/2018/11/european-parlia
ment-adopts-report-academic-freedom/

Scholars at Risk – SAR (2020). Free to Think 2020. Report. Retrieved from ht
tps://www.scholarsatrisk.org/resources/free-to-think-2020/

Telli, A. (2021). Researchers' Voices in/around/across Mobility & Time. Presentation held at the Inquiring Healing Across Screen Cultures Workshop. 1–2 June 2021. http://recuperativescreen.org/abstracts.html

Telli-Aydemir, A. & Diner, C. (2021). *Supporting Scholars in Exile: Towards Long-Term Career Path Solutions*. Academics in Solidarity Policy Brief #1. Retrieved from https://www.fu-berlin.de/en/sites/academicsinsolidarity/Policy/_1/index.html

Vatansever, A. (2020). *At the Margins of Academia: Exile, Precariousness, and Subjectivity*. Leiden: Brill.

Zapp, M., Marques, M., & Powell, J. J. W. (2018). *European Educational Research Reconstructed: Institutional change in Germany, the UK, Norway and European Union*. Oxford: Symposium Books.

South-South perspectives

Expelling and receiving scholars
Recurring purges at universities and the emergence of alternative hubs for knowledge production in Turkey

Olga Hünler

The at-risk academics' movement both inside and outside of Turkey is not a well told history. Since the beginning of the higher education system's modernization in the nineteenth century in the Ottoman Empire, educational reforms have been followed by purges of progressive scholars from universities. The consequential authoritarian interventions in universities were presented as higher education reforms by the authorities, and unfortunately, the majority of those reforms did not improve the quality of the educational system and facilities and the scientific knowledge production. Rather, they were instrumentalized to exclude liberal, progressive, or dissident scholars who did not conform to the mainstream ideology of the regime.

This chapter will follow this spiral of purges and forced replacements of scholars from the late Ottoman period to contemporary Turkey. How this cyclical model of "forced mobility" of actors in higher education shaped higher education in Turkey, as well as the repercussions and ramifications of this "academic mobility" on Turkish higher education in the core values of higher education, including institutional autonomy, academic freedom, and social responsibility of scholars to society, will be discussed accordingly. This chapter begins with a description of the first attempts at higher education reform in 1869, followed by the radical interruption in 1933 and consecutive military coups that changed the university system and repeatedly pushed the dissident scholars and critical knowledge outside of the academic system. At the end of this chapter, I discuss the specific features of academic mobility in Turkey today, as Turkey has not only become a country that pushes critical scholars out, but also has become, at the same time, a country that receives at-risk scholars from other countries, especially Syria. The limited inclusion of refugee scholars from Syria in the Turkish higher education system is addressed together with the alternative knowledge production hubs created after the massive purge of 2016.

First waves of discharges from universities: Late Ottoman Empire and the Early Republic

The history of higher education in the Ottoman period dates back to the Tanzimat period as a part of the modernization attempts of the Ottoman Empire. The first institution of higher education, known as the Darülfünun[1] opened in 1863 and closed after two years. The second Darülfünun opened in 1869 and closed in 1871. Finally, the third, Darülfünunu Sultani, which provided diplomas in engineering and law, closed in 1882 (Dölen, 2008).

The Regulation of Public Education Law (Maarif-i Umumiye Nizamnamesi) in 1869 implemented a relatively secular education system; however, the enforcement of this law took more than 10 years. The proposal of the Darülfünun provoked strong protests by the religious schools (Madrasas) and religious circles (Ege & Hagemann, 2012).

As a result of conservative protests in 1870, the first purge of progressive academics from the Darülfünun occurred (Hatiboğlu, 2000). In 1909, the first large-scale elimination of the Darülfünun began when a total of 185 professors were expelled from the university after the merger of civil and military medical schools (Bahadır, 2007). The Committee of Union and Progress (İttihat ve Terakki Cemiyeti) already had strong military relations with Germany, and it wanted to improve the educational system in line with the German system (Ege & Hagemann, 2012). In 1915, the Minister of Education (Maarif Nâzırı), Şükrü Bey, invited 20 scholars from Germany, though the German scientists were not welcomed by their Francophone Turkish colleagues (Dölen, 2008). During this period, many students were sent to Europe, especially to Germany, and their visions of a Humboldtian university shaped the Darülfünun upon their return. From 1915 to 1918, the plan to open an "Institute

---

1   According to the Encyclopedia of Islam, the name Darülfünun (Dār al-Fünūn), translates to the House of Science and is preferred to denote the modern university and to differentiate it from the madrasas which taught ulum (traditional sciences). Ege and Hagemann (2012) used House of Knowledge as translations of Darülfunun.

of German Education" did not materialize, and the German professors were forced to leave following the Armistice of Montrose (Dölen, 2010c). After their departure, the German scholars' effects on the Darülfünun disappeared very quickly. The influence of France on the Turkish education system increased at this time, and the number of invited French professors consistently increased (Dölen, 2010c).

Shortly after the proclamation of the republic in 1923, public and political debate heavily stressed the need for educational reforms to support the goals of the newly established republic. In 1924, the famous US philosopher and pedagogue John Dewey was invited to Turkey to investigate and improve Turkey's higher education system. In his report, among other things, he recommended sending successful students abroad (Yanardağ, 2017). In 1924, the Darülfünun achieved institutional autonomy, but scrutiny of the quality of education and the criticisms among the conservative circles weakened this autonomy. In 1926, French professors were invited to the science faculty to help improve its higher education system. By virtue of cultural exchange agreements with France, the first group of French professors started to work as chairs of mathematics, physics, and electro-mechanics, among other disciplines (Kadıoğlu, 2004).

The inquiry regarding the quality of the education continued, and critical columnists, intellectuals, and parliamentarians questioned the ability of the Darülfünun to nurture the young generations of the new republic. Some of these criticisms were contradictory to each other, while some groups accused the Darülfünun of not being progressive enough and lagging behind the revolutionary aspirations of the new republic; other groups, such as the intellectuals of *Kadro* magazine[2] criticized the Darülfünun as being too liberal. Meanwhile, administrators were defending themselves for being the followers and protectors of the revolution (Mazıcı, 1995). From these critiques, two main lines of criticism developed. One asserted that the Darülfünun was not capable of producing useful knowledge for fulfilling the people's needs. The other accused the institutions of being irresponsive to the revolution (Katoğlu, 2007). Ege and Haggeman (2012) argued that "[i]t

---

2   *Kadro* was an influential yet short-lived Kemalist political magazine. *Kadro* stressed the notions of anti-imperialism, independence, anti-liberalism, and statism (state control of the economy) in order to integrate itself into the Kemalist movement. According to Tanıl Bora (2017), it expected Kemalizm to become the force that awakened the East against imperialism.

seems that the decision to abolish the university had been taken before any scientific justification of the necessity for abolition had been advanced. At the ideological level, the condemnation of Darülfünun seems to have preceded any justification of the condemnation" (p. 953).

While the republican press (such as the *Kadro* magazine and *Cumhuriyet* newspaper) accused the Darülfünun professors of not defending the interest of the nation and not supporting the revolution, after several heated debates took place at the Grand National Assembly on the budget and organization of the Darülfünun, Professor Albert Malche, a professor of pedagogy at Geneva University, was invited to İstanbul to draft a detailed report on Turkish higher education in 1932 (Kadıoğlu, 2004; Tekeli, 2019). According to Dölen (2010c), Malche's first report was translated from French and carefully studied by President Atatürk and a group of senior officials. However, they decided to make radical changes in the educational system, and they abandoned Malche's report. Furthermore, Dölen (2010a) noted, Dr. Reşat Galip[3] was appointed Minister of Education on September 19, 1932 (Dölen, 2010a, p. 84). Later, Malche's second visit led to his active involvement in the Darülfünun reform of 1933. His original report on higher education had never been fully shared with the public until 1939, yet he was held accountable for the difficulties in implementing the reforms (Dölen, 2010a).

When Dr. Reşit Galip gave a lengthy statement to The Anadolu Agency of University Reform, he summarized Malche's report and criticized some professors and lecturers there. He argued that some professors prioritized their personal business operations and undermined their academic posts, did not publish influential and original scientific research, and engaged in conflicts and oppositions with their colleagues instead of sustaining unity in their ideas and aims (Dölen, 2010a). However, denigrating the purged professors as academically insufficient was not convincing; the majority of

---

3   Reşit Galip was a medical doctor and a politician. After his appointment as the Minister of Education in 1932, 92 academics were dismissed from the Darülfunun by a letter he signed in 1933 (Mazıcı, 1995). Reşit Galip emphasized the ideological and political functions of a university and defended state intervention in higher education. He proposed an Institute of Turkish Revolution, where the professors could only be those who belonged to the Turkish race (İnan, 1984). Durgun (2020) argued that "[H]e believed that the most remarkable characteristics of the university reform were that the reform was nationalist and innovative. With this arrangement, an academy at the disposal and service of politics was started" (p. 5).

the 150[4] expelled professors were able to speak and publish in several Western languages, and many of them had international publications to their credit, including monographs and articles (Mazıcı, 1995). The academic qualifications of the expelled scholars were not taken into account during the reform, and personal relations, animosities, and ideological positions were the real reasoning behind the firing and hiring of professors of the late Darülfünun (Dölen, 2010a), and this pattern was reiterated many times after this. While Özatalay (2020) has explained the purge in 1933 as a"transition from a liberal to a dirigiste economy in the wake of the Great Depression," this "transition found its ideological counterpart in a sweeping anti-liberal purge at the Darülfünun" (p. 2). Mazıcı (1995), in contrast, argued that the political ideology of *Cumhuriyet Halk Partisi* (CHP; Republican People's Party) from 1923–1945 was never oriented toward liberalism and that with the consolidation of the party-state, CHP rule eradicated any individual or organizational initiative that contradicted the official ideology.

Timur (2000) called the emergent university a product of a paradoxical period. While the Kemalist regime was becoming increasingly nationalistic, a trend influenced by similar shifts in Italy and Germany, the modernization of the university had become imperative. The void created by the expelled scholars was filled with expelled scientists with Jewish family backgrounds from Germany after the release of *Gesetz zur Wiederherstellung des Berufsbeamtentums* (Reestablished Civil Service Law), on January 30, 1933. A total of 190 persecuted scientists came to Turkey to teach at Istanbul University and modernize the newly established higher education program (Reisman, 2007; Reisman & Capar, 2007, see also Seyhan in this volume).

Integration of the foreign professors into the new university was not effortless. In his memoirs, Ernst Hirsch (1985) spoke of the lack of transparency in the secularization of the state, which was responsible for misunderstandings, tensions, and conflicts. Even though the first 10 years of İstanbul University (1933–1943) has been called its *golden age*,[5] foreign

---

4   There is a disagreement on the number of expelled professors. According to Hatiboğlu (2000), 157 of 240 professors were expelled, while Mazıcı (1995) asserted that 92 of 151 Darülfünun professors were expelled.

5   In multiple interviews conducted with the alumni of İstanbul University, the alumni referred to this first decade as "the golden age." In those interviews, this period was described as the epitome of a scientific university that has never been achieved again. The glory was attributed to Atatürk's visions and the presence of German professors (see Dölen, 2010b).

professors were criticized for not working hard enough to raise young Turkish scientists and create a scientific tradition during their tenure (see Dölen 2010b). Moreover, during the first 10 years, the income differences between foreign and Turkish professors caused bitterness, and the foreign professors faced procedural issues in their retirement that were never resolved (Dölen, 2010b; Hatiboğlu, 2000).

By the latter the part of 1930, nationalism, pan-Turkism, and pan-Turanism had become the dominant ideologies of the Turkish state. Pan-Turkism reached its peak at the beginning of 1940s, and the idea of the unity of Turks escalated with the attack of Nazi Germany on the USSR. The nationalist groups did not mind sharing their admiration of Hitler in Turkish newspapers (Bora, 2017). In 1941, a pan-Turkist committee was founded with the encouragement of Germany, and pan-Turkist parliamentarians were invited to join the cabinet (Zürcher, 2000). According to Dölen (2010c), under the given circumstances, even the slightest democratic demands, criticisms of racism, or liberal thoughts were stigmatized as communist propaganda.

Throughout the 1930s, single party regimes, such as Fascist Italy, served as alluring models, while communism, socialism, democracy, or liberalism had been discredited, though for varied reasons, by Kemalist elites (Ahmad, 1993). After the first opposition party, the Free Republican Party, which was founded upon President Kemal Atatürk's request, dissolved itself (Başaran İnce, 2015), the Republican People's Party banned all cultural and social organizations that existed following the Committee of Union and Progress Period (such as *Türk Ocakları* [Turkish Hearths], Turkish Women's Union, Women's Organization, Masonic Lodge, and liberal-socialist newspapers) (Zürcher, 2000). Finally, it was declared in the 4th Grand Congress of the Republican People's Party in 1935 that the Republic of Turkey was the first party state (Koçak, 2013). This ideological frame was not only pursued by the government and the ruling elite, but also by the academics remaining at the university after the purge in 1933. The intellectuals of the era were mobilized to spread the Kemalist ideology, especially their "modern, secular, independent Turkey imaginations" via the press and educational institutions (Zürcher, 2017, p. 182).

The second wave of purges from universities: From 1948 to 1980

Kemalism of the 1930s was called "the third period" by Bozarslan (2006, p. 32) and described as the period of becoming an autonomous ideology, encompassing six principles: secularism, nationalism, republicanism, revolutionism, étatism, and populism. After the death of Atatürk in 1938, İsmet İnönü was made the permanent party chairman and became the millî şef (national leader). In the second half of the 1940s, discontent with the economic measures that targeted the Republican People's Party's and external pressures for democratization pushed the government to allow a certain degree of liberalization (Zürcher, 2017). The postwar years were marked by the liberalization of political life, which was characterized by allowing the formation of new political parties, universal suffrage, and direct elections (Yapp & Dewdney, n.d.).

The country continued to invest in education, and the literacy rate increased incrementally (Taeuber, 1958). Prior to the new wave of purges in 1948, two new universities, Ankara University and Istanbul Technical University, were established. In 1946, universities were granted institutional autonomy. Additionally, their administrative structure and university organization were established by University Law Number 4936 (Dölen, 2010c).

Nevertheless, in the early 1940s, several scholars were arrested for their scholarly work. Scholars of Ankara University's Faculty of Languages, History, and Geography (Ankara Üniversitesi, Dil, Tarih ve Coğrafya Fakültesi, DTCF), including Pertev Naili Boratav, Adnan Cemgil, and Behice Boran, published several articles on fascism, freedom, and democracy in the new Adımlar magazine and led a vocal critique of the government (Koçak, 2007). Muzaffer Başoğlu (Sherif) published his book entitled Irk Psikolojisi (Race Psychology) in 1943 and several articles on racism and psychology. He criticized the use of Nazi inspired racial doctrines such as Turanism by Turkish nationalists and the racist idea of superior races, and the book created discontent and frustration among the racist groups. Başoğlu was arrested in 1944 with other antifascist faculty and charged with "actions inimical to the national interest" (Trotter, 1985 cited in Rusell, 2016, p. 341) and held in solitary confinement for 40 days (Russell, 2016). He later moved to the United States and continued his career at Princeton, Yale, and the University of Oklahoma, respectively. He retired from Pennsylvania State University (University Park, PA) in the 1980s (Batur, 2017).

During this time of conflict, professors Pertev Naili Boratav, Behice Boran, Niyazi Berkes, and Mediha Berkes of DTCF; five German refugee professors, Benno Landsberger, Hans Gustav Güterbock, Wolfram Eberhard, Walter Ruben, and Tibor Halasi-Kun of Ankara University; and Sadrettin Celâl Antel of İstanbul University were all picked as targets of a witch hunt initiated by right-wing groups and were pursued by the government. German professors were accused of supporting leftist faculty and of opposing racism (Dölen, 2010c). In the first expulsion attempt, the Board of Education investigated Boratav, Başoğlu, Boran, and Berkes. The dean of the DTCF authored a secret note that provided the Board of Education with the names of the professors who authored articles for the left-leaning *Görüşler Dergisi* (Görüşler Journal).[6] After an initial futile attempt, the Ankara University Senate opened an investigation. One year after that, even though the Senate decided to dismiss those professors, they applied for an appeal at the Interuniversity Council, and the Council overturned the initial verdict. However, right-wing politicians, including Prime Minister Hasan Saka, pushed for a criminal lawsuit despite all three professors being acquitted of any charges. To terminate this legal battle, the CHP government ultimately canceled the professors' tenure and expelled them from Ankara University in 1948 (Ak, 2015; Çetik, 2008; Dölen, 2010c; Öztürkmen, 2005). In addition to Boran, Boratav, Berker, and Sertel, some German professors were also fired, allegedly due to budget shortages (Hatiboğlu, 2000; Koçak, 2013). Meanwhile, professors Benno Landsberger, Hans Gustav Güterbock, Wolfram Eberhard, Walter Ruben, Georg Rohde, and Tibor Halasi-Kun[7] were named as antifascist scholars (Dölen, 2010c).

When speaking of their purge in 1948, Naili Pertev Boratav shared the following anecdote:

> In those times, there were protests against us in the faculty, calling for our deportation. This was also the time when the faculty assembly decided against us. One of those days I came across Landsberger, he was also a member of the faculty assembly. I told him, in the complaint, "Professor what has happened to us recently may also happen to you one day." And he responded, "What shall we do? There were strong charges against you."

6   Mumcu, U. (1990) 40'ların Cadı Kazanı, Tekin Yayınevi, (pp. 104–105) as cited in Dölen, 2010c.

7   Ultimately, Georg Rohde was discharged instead of Tibor Halasi-Kunt, and the reason for the discharge of those five foreign professors was never clarified (Dölen, 2010c).

Nevertheless, together with us, 25 émigré professors, including Landsberger and Güterbock were also deported.[8]

According to the university law passed in 1946, only public servants could apply for the associate professorship (habilitation) exam. This meant that only Turkish nationals could be candidates for these positions, thereby obstructing the German professors' career growth in favor of the nationalist ideology of the state (Hatiboğlu, 2000). Even though German professors had been allowed to apply for Turkish nationality, this decision had many downsides, including losing 75% of their salaries (Hatiboğlu, 2000). Anti-Semitic sentiment in Turkey fostered by German propaganda and Pan-Turkism, conflicted with Turkish and pro-Nazi German academics as well as with the Turkish government, and a hostile political environment in Turkey influenced the departure of German professors to America and Europe in the 1940s, once their work contracts were terminated or ended, or when they found new opportunities (Tomenendal et al., 2010).

At the end of the one-party rule, the Democrat Party (DP), which was the third legal oppositional party as well as the first party to form a government, de-seated the Republican People's Party (formerly known as the People's Party) by winning the 1950 national elections. In the first years of the DP, the government had positive relations with the universities. However, this fragile balance started to weaken when the DP introduced a change to the University Law which criminalized professors' political activities and comments (Dölen, 2010c). Their regulations were a direct threat to university autonomy, and the deposal of the Dean of the Faculty of Political Science, Turhan Feyzioğlu, caused substantial discontent among academics and intellectuals (Dölen, 2010c; Hatiboğlu, 2000). However, the ephemeral reign of the DP was terminated on May 27, 1960[9] by the military's National Unity Committee, and the military regime lasted for the next 18 months. Strangely, the new revolutionary *zeitgeist* of the regime hit the universities first when the National

---

8   Çetik. M. (1998). 1948 DTCF Tasfiyesi ve P.N. Boratav'ın Müdafaası [The 1948 purge in DTCF and the defense of P.N.Boratav], Üniversitede Cadı Kazanı, p.198, as cited in Tomenendal et al., 2010.

9   The 1960 Turkish coup d'état was the first, but not the last military coup in the Turkish Republic. After the coup, 592 Democrat Party members were put on trial in the military courts and three of them, Adnan Menderes (Prime Minister), Fatin Rüştü Zorlu (Minister of Foreign Affairs), and Hasan Polatkan (Minister of Treasury) were executed (Özdemir, 2007).

Unity Committee changed the University Law again and dismissed 147 academics from universities (Dölen, 2010c; Timur, 2000). Ironically, some of the dismissed professors and professors-in-ordinary were among the new faculty hired at İstanbul University after the 1933 purge (Dölen, 2010c).

Third wave of purges after the 1980 coup and the privatization of higher education

After the first and second waves of purges in 1933 and 1960, respectively, another massive expulsion from universities occurred after the 1980 coup d'état. Some research assistants and lecturers were dismissed following the 1971 military coup, yet the purge was relatively minor.

The end of the 1970s was marked by polarization and frustration of the Turkish people and escalated the economic and political instability of the country. Terror attacks targeted left-leaning academics and intellectuals as well as Alevis, a minority among the Sunni majority, often denounced as "communists" by the right-wing Grey Wolves[10] (Ahmad, 1993).

On September 12, 1980, armed forces seized political power, dissolved the parliament, unseated the cabinet, rescinded immunity of the members of the parliament, and suspended all political parties and two trade unions (Zürcher, 2000). Not immediately, surprisingly, the martial administration started to revise the University Law, and they established the Council of Higher Education (CoHE, in Turkish, its acronym is YÖK) in a capacity defined by Articles 130 and 131 of the 1982 constitution in order to control the universities. "Universities were thus being disciplined for not addressing terror and violence, and not keeping out of political struggles" (Güvenç, 1990, p. 90).

The military regime utilized multiple methods for their purge. They reorganized the old universities and opened new ones. They forced the

---

10   The youth organization of the Nationalist Action Party (Milliyetçi Hareket Partisi) officially called the Hearths of Ideal (Ülkü Ocakları) and the members named themselves Greys Wolves. They started threatening leftist students, teachers, publicists, and booksellers. "The Grey Wolves received paramilitary training in specially designed camps and, like Hitler's SS, their mission was to conquer the streets (and the campuses) on the left" (Zürcher, 2017, p. 260).

faculty to work on a rotational system.[11] Scholars who resisted the rotational system and scholars who were known as communists, or the ones targeted in personal conflicts of interest, were dismissed by Martial Law No. 1402, which was established in the aftermath of the military coup. Some of the academic personnel were dismissed by the rector's orders, while others were either banished or threatened to be purged by means of this law (Dölen, 2010c; Güvenç, 1990). Additionally, some professors resigned to protest the situation (Hatiboğlu, 2000) or resigned because of the mobbing of the university administrations and colleagues (Dölen, 2010c). According to Dölen (2010c), we do not know the exact number of purged academics since some of them refused to return to the universities even after the State Council decisions to allow their return. Hatiboğlu posited that the number of purged academics might have ranged from 1,200 to 1,300.

Universities were restructured once again by Higher Education Law No. 2457. First of all, the new law organized every education institute (including institutes and vocational schools regulated by the Ministry of Education) under the YÖK, centralized the educational system (Birler, 2012), and revoked the universities' autonomy once again. The law mandated the appointment of administrative personnel, such as rectors, deans, and department chairs, instead of elections and regulations of academic promotions by the YÖK. Higher education was also restructured to serve the needs of the rising private sector. A two-tier system was introduced, and a few top-ranking universities started to instruct in English to support the newly emerging "managerial and technocratic class" (Ahmad, 1993, p. 210).

Finally, with the new regulation, the establishment of the nonprofit foundation universities was enabled, and higher education fees were introduced even for the public universities (Katoğlu, 2007). In the 1990s, the number of private universities in Turkey started to increase rapidly.[12] In fact, since 1992, 74 private universities have opened in Turkey. The increase of the foundation of universities after 2002 caused an increase

---

11   It was a mandatory system to force academics from established universities at the center to new universities at the periphery. These rotations were a punishment of the dissident academics rather than an opportunity to improve the quality of the education in newly established universities (Versan, 1989).

12   According to the statistics provided by YÖK, there were 74 private/foundation universities, 129 state universities, and four private vocational schools in Turkey by 2020 (see https://www.yok.gov.tr/universiteler/universitelerimiz).

in the commodification of the universities, and they also began to be dispersed to smaller cities, such as Gaziantep, Kayseri, Konya, Trabzon, Samsun, Bursa, and Antalya (Birler, 2012). Higher education's marketization alters "universities into corporations, faculties into departments, university presidents into managers, and academicians into workers" wrote Önal (2012, p. 136). Vatansever and Yalçın (2015) described this academic milieu as a toxic environment that permeates insecure and trivialized intellectual production and creates involuntarily nomadic and precarious academics who have been threatened with unemployment and dismissals by the administeration after the intense privatization of higher education. As Biner (2019) described, "[W]ith the ascension of AKP rule in 2002, neoliberal policies aggressively and rapidly penetrated the system of higher education, and Erdoğan set about imposing stricter control over the YÖK and the universities" (p. 20). Vatansever (2018) highlighted this impermeable knot of authoritarianization and privatization as follows: "the ongoing witch-hunt in the universities adds a political dimension to the hitherto economic precarization of the academic labor force, and should be seen as part of a wider, distinctly neo-liberal attempt on the part of the state to eradicate rational agency. By eliminating qualified oppositional cadres en masse on false accusations, the government is implementing a systematic deinstitutionalization of intellectual production" (p. 5).

The last wave of discharges from universities: the Peace Petition and failed coup attempt of July 2016

Turkish higher education did not fully recover from the damage created by the junta's regime that was established after the 1980 military coup. Universities could not attain institutional autonomy or achieve freedom to teach and research fully, and during the 1980s, the YÖK shaped academic life in Turkey. Since the relatively quiet years of the 1990s and 2000s, sporadic attacks on individual scholars[13] or certain research topics[14] have escalated and been extended to all higher education (Baser et al., 2017).

A couple of months before the coup attempt in July 2016, a group of academics calling themselves Academics for Peace (AfP) released an open letter to the Turkish state demanding that they stop using violence and breaching human rights in the Kurdish regions of Turkey. They also demanded that they resume peace negations. When the petition entitled "We will not be a party to this crime!" was shared with the public, the initial signatures totaled 1,128, with academics inside and outside of Turkey signing it. President Erdoğan reacted ferociously upon its release, and he accused the academics of treason and labeled the signatories as "ignorant" and "so-called intellectuals" and of being a "fifth column" and "dark."[15] The YÖK immediately responded to Erdoğan's call on the judiciary and university administrations by issuing a statement that claimed that the Peace Petition could not be accepted as a protected exercise of academic freedom.[16] Immediately following this, 30 academics were detained, their houses were raided by anti-terror police, and the number of detentions increased to 70 (Abbas & Zalta, 2017). Eighty-nine academics were dismissed with or without disciplinary investigations; some academics faced disciplinary actions including warnings, demotions, reprimands, or suspensions (Abbas & Zalta 2017; Baser et al., 2017).

Not surprisingly, nationalist and conservative media were encouraged by the president's intimidating speech, and they too targeted the academics. Unfortunately, some of the university senates and chancellors jumped on the bandwagon with the media, and they issued several statements denouncing their colleagues as "supporters of terrorism," "slanderous," "so-called academics," or "vile" (Sözeri, 2016; Tekin, 2019).

---

13 Such as İsmail Beşikçi, a sociologist who studied the Kurdish question and has been imprisoned for 17 years, or Büşra Ersanlı, an eminent political scientist who gave lectures at the academy of BDP, the Kurdish political party at the time, who was imprisoned for nine months in 2011.

14 Such as the Armenian genocide, the Dersim massacre, or the Kurdish question.

15 See https://bianet.org/bianet/ifade-ozgurlugu/171150-erdogan-dan-akademisyenlere-d aga-ciksinlar-veya-hendek-kazsinlar?bia_source=facebook&utm_source=dlvr.it&utm_ medium=facebook

16 Very unexpectedly after the defamation campaign started, almost 1,000 academics signed the petition, and the total number exceeded 2,000 (https://barisicinakademisy enler.net/node/1).

Eight hundred and twenty-two academics have stood trial since December 5, 2017. Of the 108 academics whose cases were concluded, all were sentenced to serve from 1 year and 3 months to 3 years in prison. Twelve were convicted. By January 2021, there were still 91 ongoing trials despite the Constitutional Court's decision that the penalization of Academics for Peace on charges of "propagandizing for a terrorist organization" violated their freedom of expression. Even though more than 100 academics were dismissed, forced to resign, or retired because of signing the Peace Petition, the real purge occurred after the July 2016 coup attempt.

The consecutive statuary decrees caused permanent changes to the institutions, including universities (TİHV Akademi, 2018). They also destroyed what remained of universities' institutional autonomy (Taştan et al., 2020). During this prolonged state of emergency, the government abolished the rectorate elections, and the president licensed himself to appoint the rectors. Considering the current political climate of the country, rectorate appointments are not expected to reflect the academic merit[17] of the candidates or their administrative expertise. Rather, appointments are more so based on the degree of ideological agreement with the government. Similarly, universities' authority to conduct disciplinary procedures with their academic staff has been delegated to the Chairperson of the Council of Higher Education. According to the 2016–2017 report of the Science Academy, "These new disciplinary regulations are in continuation of the repressive tradition established in our country in 1980. On the one hand, universities' power to perform disciplinary investigations as independent legal entities is partially transferred to the Council of Higher Education, on the other hand, academics' freedom of expression is limited in an unconstitutional manner" (The Science Academy, 2017, p. 10).

With the consecutive statuary decrees, in addition to thousands of civil servants such as teachers, nurses, doctors, and technicians, 1,427 administrative personnel, and 6,081 academics (2,493 from social sciences

---

17   In January 2021, President Erdoğan appointed Professor Melih Bulu as a rector of one of Turkey's most renowned institute, Boğaziçi University. The appointed rector was accused of plagiarism (https://scienceintegritydigest.com/2021/01/07/newly-appointe d-bogazici-university-rector-accused-of-plagiarism/), and his merits were questioned widely by academics and students of the university (https://m.bianet.org/english/hum an-rights/238559-academics-call-on-appointed-rector-to-resign-keep-your-hands-off-my-student).

and humanities, 1,886 from health sciences, and 828 from engineering departments, 81 from sports and arts, and the academic fields of 98 academics that were not mentioned) were dismissed from universities without explanation or due process of law (TİHV Akademi, 2018). All 1,576 deans from all public and private universities were obliged to resign from their posts (Aydın et al., 2021) and public servants including the academics were not allowed to travel abroad in the following days of the coup attempt and later they were only allowed to travel with a certificate of clearance from their working institutions until November 15, 2016. Meanwhile, students also suffered from decree laws. While 300 graduate students studying abroad were expelled and lost their scholarships, 15 universities were closed (TİHV Akademi, 2018), and students were compelled to register at other universities without an informed decision and mostly with a status of "special student," some having to study in segregated classes (Namer et al., 2018).

Even though this last in the wave of purges has been the most extensive one, which affected thousands of academics and students, like the previous ones, this purge was also caused by the state's desire to implement its political (as well as social and economic) agenda to consolidate its domination and influence over higher education. This intervention has gravely damaged the remaining autonomy of the higher education institutions and severely impacted the ability to pursue the scientific and academic endeavors of the scholars.

Syrian (refugee) scholars in Turkey

The failed coup attempt created an immense opportunity to restructure the Turkish higher education system and purge the thousands of undesirable academics from the universities. Some of the purged academics left Turkey to work in foreign universities. Even though it is difficult to ascertain the exact number of exiled academics from Turkey since 2016, Turkish scholars constitute the largest group of applicants to the SAR network, and from January 2016 to October 2019, SAR received more than 1,000 applications from Turkey.[18] Several initiatives including the global Scholar Rescue Fund, the Philipp Schwartz Initiative of the Alexander von Humboldt

---

18   See https://www.irishtimes.com/news/education/european-alliance-for-academics-at-risk-to-be-based-out-of-maynooth-university-1.4049383

Foundation in Germany, Norwegian institutions through the Students at Risk program (Tekdemir et al., 2018), Academy in Exile Fellowship Program in Germany, Academic Freedom Program of the Einstein Foundation Berlin, and Programme d'aide à l'Accueil en Urgence des Scientifiques en Exil (National program for the urgent aid and reception of scientists in exile, PAUSE) in France, among other programs, host the growing number of Turkish scholars at risk.

While Turkish universities were losing thousands of academics, refugee academics and students primarily from Syria were struggling to find themselves an opportunity to continue their research and education in Turkey. According to the statistics collected by the YÖK, by 2017, 392 Syrian academics (including professors, associate and assistant professors, lecturers, research and teaching assistants, and educational planners) were employed in Turkish higher education institutions. This number constituted 14% of all foreign academic personnel but only 0.2% of the total number of academics (Erdoğan et al., 2017).

In 2018, a group of professors, most of whom were the rectors of the Turkish universities, in cooperation with the YÖK and under the patronage of the president, started a project entitled the Preservation of Academic Heritage in the Middle East Project. On their webpage,[19] they introduced the project as follows:

> This project was prepared to support the dreams of the students and scientists, who were forced to flee their countries due to war conditions, in order to ensure that they resume their academic lives, and to make their voices heard.
>
> Talking about the incomplete academic lives and destroyed science centers of the Middle East on different sides of the world and keeping them alive is an academic heritage that will be passed on to the next generations.
>
> What is lost does not belong only to the heritage of the Middle East but to all humanity as well. The maintenance of scientific environment in the Middle East by promoting scientists and students will ensure the common future of humanity and the reconstruction of the ruined regions.

Since the project started in 2018, its participants have appeared mainly on pro-government mediums, such as the Turkish state-run Yunus Emre

---

19   See http://www.akademikmiras.org

Institute in London.[20] The project panels co-organized by the YÖK at the School of Oriental and African Studies (SOAS) University of London, and the University of Duisburg-Essen, Germany were protested by scholars and activists, and the panel at SOAS was subsequently cancelled.[21] The speakers of those panels were rectors and deans of the Turkish universities as well as members of YÖK, politicians, and representatives of pro-government foundations. Refugee scholars were not provided an opportunity to speak on their own behalf and raise their voices.

On the web page of this project, short videos of refugee students and scholars disclosing their experiences as well as their "gratitude to Turkey,"[22] which were produced with dramatic cinematography were shared. On this web page, stakeholders did not publish any road map or tangible plan for integrating refugee scholars and students into the higher education system in Turkey; instead, they seemed to prepare propaganda materials that attempted to restore the compromised reputation of the Turkish educational system and the YÖK.

While the Preservation of Academic Heritage in the Middle East Project attempted to promote itself at international events, Syrian academics experienced several difficulties in Turkey, including a lack of professional opportunities, economic hardships, and discrimination (see Ghazzoul in this volume). In a study conducted by a team of researchers from the University of Cambridge Faculty of Education and Syrian Co-researchers published by the Council for At-Risk Academics (CARA), researchers interviewed nineteen displaced male academics residing in Turkey at the time of the research. Some participants reported that they had to teach Arabic, regardless of their academic field, to survive in a foreign country with limited academic job market opportunities. One participant shared his experience:

> Now I am teaching Arabic in [name of the Faculty]. It is not my specialization, but I can get enough money for me and my children. I am refreshing my English and trying to publish in my specialization, and as you know, my job is merely a way to get money. It is like I am beginning again, to be honest."
> (Interviewee 19) (Dillabough et al., 2019, p. 81)

---

20   See https://ahvalnews.com/academics-peace/soas-backs-out-panel-discussion-after-ba cklash-turkish-academics
21   See https://academicboycottofturkey.wordpress.com/news/
22   See https://youtu.be/heYcsZMmDOY

Parkinson et al. (2020) observed that the refugee academics whom they interviewed in Turkey had difficulty finding jobs related to their academic interests, and sometimes they felt obligated to work outside of academics, mainly in unskilled jobs. Teaching courses outside of their expertise, not having access to scientific resources, and the lack of professional and academic support have made refugee scholars unsettled and frustrated. Correspondingly, more than 80% of Syrian academics employed at Turkish universities have a background in theology or religious studies, and they found more permanent working opportunities at the theology faculties or religious vocational schools. A significant number of Syrian scholars, probably from other disciplines, are assumed to have left Turkey for third countries (Erdoğan et al., 2017). Despite the Turkish government's resistance to collaborating with international actors, there has been support from non-governmental organizations (NGOs), working within the unclear and continuously shifting legal space in Turkey (Watenpaugh et al., 2014). However, Syrian scholars have not been content with the amount of support they have received from international organizations (Abdo, 2015; Parkinson et al., 2020; Watenpaugh et al., 2014). Moreover, neither the quality nor the amount of support refugee academics receive in Turkey has seemed satisfactory to them. The cancelation of work contracts caused Syrian academics to lose their already precarious, low-paid jobs in higher education. McLaughlin and colleagues (2020, p.7) quoted two interviewees: "At one new university, the hopes that working would mean being professionally recognized and legitimated so central to testimonies of professional integrity turned out 'not to be true' (Karam) and 'unfortunately, I also saw the criminals and thieves [in the new HE context]' (Jamal)."

Because they were disconnected from their academic and professional networks, lacked new research collaboration opportunities, and had problems with their legal status (Parkinson et al., 2020; Watenpaugh et al., 2014), in addition to facing extensive discrimination and racism directed toward refugees in Turkey (Doğanay & Çoban Keneş, 2016; Şimşek, 2017), Syrian academics experienced loss of professional and academic career pursuits (Dillabough et al., 2019). The case of Syrian scholars in Turkey well illustrates the spiral of continuous dismissals. While the employment of refugee scholars from Syria is presented by governmental institutions as a humanitarian act to preserve the heritage of the Middle East, these scholars are placed on the periphery of the university. Once again, the Turkish higher educational system pushed a significant number of researchers outside the universities, and this time not so much as the result of a systematic purge of ideological university reform, but because of disregard or indifference of the government toward a refugee scientist.

Solidarity Academies

Having described the different purges and the half-hearted integration of Syrian refugee scholars, it remains important to observe what happens outside the university. The purges do in fact transform the academic milieu, and the scholars affected by the purges are not only fleeing the country but reorganizing themselves outside the university. While thousands of academics were forced to leave their posts at universities, and the academics who remained in the universities complained about self-censorship and losing their calling for doing research and motivation for research and teaching (Aktas et al., 2019), the critical academic discussions moved to another medium, solidarity academies. Solidarity Academies is an umbrella term used to define several initiatives, such as *Kocaeli Dayanışma Akademisi* (Kocaeli Solidarity Academy, KODA), *İzmir Dayanışma Akademisi* (İzmir Solidarity Academy, İDA), *Ankara Dayanışma Akademisi* (Ankara Solidarity Academy, ADA), *İnsan Hakları Okulu* (The School of Human Rights), and *BirAraDA Dernek* (Association for Science, Art, Education, Research and Solidarity), which create hubs for producing scientific and academic knowledge outside the universities. They were established in many big cities in Turkey, and they have organized public lectures, seminars, summer and winter schools, and workshops. During the pandemic, solidarity academies pursued their activities online like many other associations. Tuğrul and Deniz (2019) described participants of solidarity academies as a core group, a stable group of academics, most of whom were dismissed from their tenure, and other less regular participants, who differ significantly in their backgrounds. The core members are primarily signatories of the Peace Petition, and many of the solidarity academies described themselves as organized horizontally and collaboratively.

Tutkal (2020, p. 5) summarized the aims of the solidarity academies as follows:

> ... to relate academic knowledge production to the prioritization of peace, nonviolence and justice in the socio-political sphere, to continue such

knowledge production processes in the non-university spheres, and to maintain their relation with the dare-to knowledge that requires courage in producing and sharing knowledge, prioritizing peace vis-a-vis the authoritarian structures [and thus] to produce and share knowledge with reference to equality, freedom, and solidarity that are excluded from the university sites.

Even though they emerged as a response to massive purges following the coup attempt in 2017, the operations, organizations, or focuses of the solidarity academies are not identical to each other (see Özgür in this volume). Each one provided a distinct contribution to the field. For instance, while the Mersin-based *Kültürhane* (Culture House) works as a non-profit social space also functioning as a public library, *Aramızda* defines its mission as "conducting research on gender and gender equality, provision of training activities, and contribution to rising awareness by sharing accumulated knowledge and experience" on their webpage.[23] In addition, TİHV Akademi, which was initiated by the academics purged by the statuary decrees from the three universities in İzmir, focuses mainly on documenting the rights violations experienced by scholars and human rights activists. They publish regular reports and bulletins to cover rights violations. Also, Off-University was established in Berlin by and for persecuted academics, who were dismissed or forced to resign from their jobs or were prosecuted or imprisoned, as well as their colleagues and friends. Off-University primarily aims to provide an opportunity to teach online and earn an income for persecuted academics of the world who can no longer work at a university and are criminalized for their opinions. Erdem and Akın (2019) described each solidarity academy as "conceptually unique and embedded in its own distinct local context" (p. 150). For example, Ankara Street Academy started teaching in Kuğulu Park[24] with a lecture entitled "Hegemony and Counter Hegemony." Unlike other academies, their audience was the people living in the neighborhood, not necessarily university students, and their aim was to offer academic knowledge to the people while choosing the topics relevant to current debates in Turkish society (Aktas et al., 2020).

Mollona and colleagues (2020) explored three popular initiatives: the *Bachilleratos Populares* in Argentina, the Landless Movement in Brazil, and

---

23   See https://aramizda.org.tr/index.php/en/main-page/
24   Kuğulu Park is a centrally located public park in the Çankaya district of Ankara, Turkey.

Kampüssüzler in Turkey, and they identified that despite their differences, they are strongly tied to feminist, anti-capitalist, and decolonial struggles. Despite the principal or practical differences among the solidarity academies, Bakırezer and Koçak (2017) pointed out three common dimensions where dismissed academics carry on their efforts: "1) recovering the lost position and return to the university; 2) creation of an alternative and democratic academic organization by restructuring academic institutions and, finally; 3) democratization of the country as the compulsory external political condition in order to have a positive academic environment" (Bakirezer & Koçak, 2017, as cited in Tuğrul & Deniz, 2019, p. 493). Aktoprak (2020, p. 22), a signatory of Peace Petition and a purged academic with a statuary decree, defined the demanding work of the dismissed academics as follows:

> We are now trying to prove that we are "no different from others," that we are still academics and do not need four walls. But this effort is no longer an effort that anchors the past. We are aware that we now need to separate our ties with the university we were purged into to ensure continuity. Not because we get the gate, but because the university is no longer a university under the authoritarian regime. [Separation] is to pursue the university, the academy. But just as the person who went through the colonial experience is not the same, we are no longer the same person, the same scholar [author's translation].

As Erdem and Akın (2019) argued, solidarity academies could benefit from their frustration with the current situation of the education system and support from civil society. For now, nobody knows when and how dismissed academics can return to their universities, or whether they would be willing to return to institutions that purged them and to colleagues who did not defend them. It is also unknown to what extent they would be willing to struggle to change or powerful enough to change the instrumentalization of universities, the less than mediocre research and education quality, and the diminished academic freedom and autonomy.

Conclusion

As Pherali (2020) acknowledged, a "mass displacement of academics places a huge cost on the country of their origin not only in terms of the loss of human capital but also the destruction of the intellectual life of the entire community"

(p. 90). Since the very beginning of the establishment of Turkish universities, academics were excluded from academic knowledge production by forces with political agendas. With every purge, political elites desired to redesign higher education according to their needs. Unfortunately, those restructuring and redesigning efforts have damaged the institutional autonomy, academic freedom, and quality of scientific production and education in Turkish universities. While four Turkish universities had ranked among the top 200 institutions globally in the Times Higher Education's World University Rankings 2014–2015, in 2021, after the purge of thousands of academics and complete destruction of the remaining institutional autonomy and academic freedom, none of the Turkish universities were even ranked in the top 400 in these rankings.[25] While the public and private universities are losing their edge and popularity in the international higher education setting, the transformative role of solidarity academies becomes more visible on a day-to-day basis. They have hosted unorthodox research topics and critical perspectives that are marginalized in mainstream academia, such as the history of labor, resistance, commoning, gender and LGBTIQ, and workplace homicides. Furthermore, since their establishment, members of the solidarity academies have become popular expert speakers at events on academic freedoms, while their documentations are extensively and internationally used by their peers.

Academics in Turkey have endured rights violations and oppression because of their research or their extramural activities since the establishment of higher education in modern Turkey. University reforms that are followed by massive purges (or vice versa) of the critically acclaimed scholars from the universities show a nearly cyclical pattern: in every decade or two, universities suffer from shockwaves of anti-democratic intrusions. The last radical intrusion caused the largest dismissal of the academics and caused irreversible damage to higher education, at least in the short term. However, this time, alternative hubs of knowledge production have flourished, and they have given the long-lost autonomy and academic freedom back to the researchers with limited resources and the dismissed academics pursuing their scientific endeavor and sharing their vast knowledge with people without economic, social, or political barriers.

---

25   See https://www.timeshighereducation.com/world-university-rankings/2021/world-ran king#!/page/0/length/25/locations/TR/sort_by/rank/sort_order/asc/cols/stats

References

Abbas, T., & Zalta, A. (2017). 'You cannot talk about academic freedom in such an oppressive environment': perceptions of the We Will Not Be a Party to This Crime! petition signatories. *Turkish Studies, 18*(4), 624–643.

Abdo, W. (2015, February 6). To be a Syrian professor: recipe for tragedy. *Al-Fanah News and Opinion About Higher Education*. Retrieved from https://www.al-fanarmedia.org/2015/02/syrian-professor-recipe-tragedy/

Ahmad, F. (1993). *The making of modern Turkey (Vol. 264)*. London: Routledge.

Ak, G. (2015). Türk düşünce hayatinda Mediha Esenel (Berkes) ve 1948 DTCF tasfiyeleri Iilişkisi üzerine bir Iinceleme. *Journal of Modern Turkish History Studies / Çagdas Türkiye Tarihi Arastirmalari Dergisi, 15*(30), 251–293.

Aktas, V., Nilsson, M., & Borell, K. (2019). Social scientists under threat: Resistance and self-censorship in Turkish academia. *British Journal of Educational Studies, 67*(2), 169–186, DOI: 10.1080/00071005.2018.1502872

Aktas, V., Nilsson, M., Borell, K., & Persson, R. S. (2020). Taking to the streets: A study of the street academy in Ankara. *British Journal of Educational Studies, 68*(3), 365–388.

Aktoprak, E. (2020). Saflara çekilmeyerek sınırlara direnmek. Yeni Bir İnsan Hakları Hareketine Doğru, Küresel İnsan Hakları Krizi Karşısında Ne Yapmalı? Symposium Proceedings, October 3–7, 2020, TIHV.

Aydin H., Mak V., & Andrews K. (2021). Academic Freedom and Living in Exile: Experiences of Academics in Turkey. In H. Aydin & W. Langley (Eds.), *Human Rights in Turkey. Philosophy and Politics - Critical Explorations*. Vol. 15. (pp. 339–363). Cham: Springer. https://doi.org/10.1007/978-3-030-57476-5_15

Bahadır, O. (2007). 1933 Üniversite Reformu Niçin Yapıldı? In N.K. Aras, E. Dölen, & O. Bahadır (Eds.), *Türkiyede Üniversite Anlayışının Gelişimi (1861–1961)*. (pp. 52–85). Ankara: Türkiye Bilimler Akademisi Yayınları, sıra 15.

Baser, B., Akgönül, S., & Öztürk, A. E. (2017). "Academics for Peace" in Turkey: a case of criminalising dissent and critical thought via counterterrorism policy. *Critical Studies on Terrorism, 10*(2), 274–296.

Başaran İnce, G. (2015). The Free Republican Party in the political cartoons of the 1930s. *New Perspectives on Turkey, 53*, 93–136. doi:10.1017/npt.2015.20

Batur, S. (2017). Muzaffer Sherif in FBI files. In A. Dost-Gozkan & S. Sönmez Keith (Eds.), *Norms, groups, conflict, and social change: rediscovering Muzafer Sherif's psychology*. London/New York: Routledge.

Biner, Z. Ö. (2019). Precarious solidarities: 'poisonous knowledge'and the Academics for Peace in times of authoritarianism. *Social Anthropology, 27,* 15-32.

Birler, Ö. (2012). Neoliberalization and Foundation Universities in Turkey. In K. İnala & G. Akkaymak (Eds.), *Neoliberal transformation of education in Turkey: Political and ideological analysis of educational reforms in the age of the AKP,* (pp. 139-150). New York: Palgrave MacMillan.

Bora, T. (2017). *Cereyanlar: Türkiye'de siyasi ideolojiler.* İletişim Yayınları.

Bozarslan H. (2006). Kemalism, westernization and anti-liberalism. In H.L. Kieser (Ed.), *Turkey beyond nationalism: Towards post-nationalist identities.* (pp. 28-37). London: I.B.Tauris.

Çetik, M. (2008). Üniversitede Cadı Avı: 1948 DTCF Tasfiyesi ve P. N. Boratav'ın Müdafaası. Ankara: Dipnot.

Dillabough, J., Fimyar, O., McLaughlin, C., Al Azmeh, Z., Jebril, M., Abdulhafiz, A. H., ... & Shaban, F. (2019). *Syrian higher education post-2011: Immediate and future challenges.* London: CARA (Council for At-Risk Academics).

Doğanay, Ü., & Çoban Keneş, H. (2016). Yazılı Basında Suriyeli 'Mülteciler': Ayrımcı Söylemlerin Rasyonel ve Duygusal Gerekçelerinin İnşası. *Mülkiye Dergisi, 40*(1), 143–184. https://dergipark.org.tr/en/pub/mulkiye/issue/374 12/432816.

Dölen, E. (2008). II. Meşrutiyet Döneminde Darülfünun. *Osmanlı Bilimi Araştırmaları, 10*(1), 1–46.

Dölen, E. (2010a). *Türkiye Üniversite Tarihi 3, Darülfünun'dan Üniversiteye Geçiş Tasfiye ve Yeni Kadrolar.* İstanbul: İstanbul Bilgi Üniversitesi Yayınları.

Dölen, E. (2010b). *Türkiye Üniversite Tarihi 4, İstanbul Üniversitesi 1933–1946,* 4. Cilt. Istanbul: İstanbul Bilgi Üniversitesi Yayınları.

Dölen, E. (2010c). *Türkiye'de Üniversite Tarihi 5, Özerk Üniversite Dönemi.* İstanbul: İstanbul Bilgi Üniversitesi Yayınları.

Durgun, Ş. (2020). Universities in Turkey: Changing politics and science. *Journal of Public Affairs, 20*(4), e2184.

Ege, R., & Hagemann, H. (2012). The modernisation of the Turkish University after 1933: The contributions of refugees from Nazism. *The European Journal of the History of Economic Thought, 19*(6), 944–975.

Erdem, E., & Akın, K. (2019). Emergent repertoires of resistance and commoning in higher education: The solidarity academies movement in Turkey. *South Atlantic Quarterly, 118*(1), 145–163.

Erdoğan, M. M., Erdoğan, A., & Yavcan, B. (2017). Elite dialogue: Türkiye'deki Suriyeli akademisyen ve üniversite öğrencilerinin durumu, sorunları ve beklentileri araştırması. Retrieved from https://igamder.org/uploads/bel geler/ELITE-DIALOGUE-01102017-TR.pdf

Güvenç, B. (1990). Turkish Higher Education in Transition from a Formal Autonomy to Academic Freedom via State Corporatism. *Beitrage zur Hochschulforschung*, 1(2), 89–100.

Hatiboğlu, M.T. (2000). *Türkiye Üniversite Tarihi*. Selvi Yayınevi.

Hirsch, E. (1985). *Hatıralarım*. Ankara: Türkiye İş Bankası.

Kadıoğlu, S.İ. (2004). 1933 Üniversite Reformu Hakkında Bir Bibliyografya Denemesi. *Türkiye Araştırmaları Literatür Dergisi 4*, 471–492.

Katoğlu, (2007). Cumhuriyet Türkiyesi'nde eğitim, kültür, sanat. In M. Tunçay, C. Koçak, H. Özdemir, K. Boratav, S. Hilav, M. Katoğlu, & A. Ödekan (Eds.), *Türkiye Tarihi. Çağdaş Türkiye (1908-1980) Cilt 4* (pp. 393–501). Cem Yayınevi.

Koçak, C. (2007). Siyasal tarih in. In M. Tunçay, C. Koçak, H. Özdemir, K. Boratav, S. Hilav, M. Katoğlu & A. Ödekan (Eds.), *Türkiye Tarihi. Çağdaş Türkiye (1908–1980) Cilt 4*. (pp. 393–501). Cem Yayınevi.

Koçak, C. (2013). *Rejim Krizi Türkiye'de İki Partili Siyâsi Sistemin Kuruluş Yılları (1945–1950) Cilt 3*. İletişim Yayınları.

Mazıcı, N. (1995). Öncesi ve Sonrasıyla 1933 Üniversite Reformu. *İletişim*, 76, 56–70.

McLaughlin, C., Dillabough, J., Fimyar, O., Al Azmeh, Z., Abdullateef, S., Abedtalas, M., … & Shaban, F. (2020). Testimonies of Syrian academic displacement post-2011: Time, place and the agentic self. *International Journal of Educational Research Open*, 1, 100003.

Mollona, M., Tarlau, R., Blaustein, L.A., Sualp, Z.T., & Mariano, A. (2020). Popular education and knowledge in commons. Institute of Radical Imagination. The School of Mutation - Radical Pedagogy.

Namer, Y., Düzen, E., Hünler, O., Uysal, M., Duman, E., & Hasateş, M. (2018, July 2–4). Narratives of being 'left-behind': Students' accounts of academic displacement. Paper presented at the 2018 Annual IMISCOE Conference, Barcelona, Spain.

Önal, N.E. (2012). The Marketization of Higher Education in Turkey (2002–2011). In K. İnal & G. Akkaymak (Eds.), *Neoliberal transformation of education in Turkey: Political and ideological analysis of educational reforms in the age of the AKP,* (pp. 125–138). New York: Palgrave MacMillan.

Özatalay, C. (2020). Purge, Exile, and Resistance: Rethinking the Conflict of the Faculties through the Case of Academics for Peace in Turkey. *European Journal of Turkish Studies. Social Sciences on Contemporary Turkey*, (30). https://doi.org/10.4000/ejts.6746

Öztürkmen, A. (2005). Folklore on trial: Pertev Naili Boratav and the denationalization of Turkish folklore. *Journal of folklore research*, 42(2), 185–216.

Parkinson, T., McDonald, K., & Quinlan, K. M. (2020). Reconceptualising academic development as community development: lessons from working with Syrian academics in exile. *Higher Education*, 79(2), 183–201.

Pherali, T. (2020). 'My life as a second-class human being': Experiences of a refugee academic. *Education and Conflict Review*, (3), 87–97.

Reisman, A. (2007). Jewish Refugees from Nazism, Albert Einstein, and the Modernization of Higher Education in Turkey (1933–1945). *Aleph*, 253–281.

Reisman, A., & Capar, İ. (2007) The German-Speaking Diaspora in Turkey: Exiles from Nazism as Architects of Modern Turkish Education (1933–1945). *Diaspora, Indigenous, and Minority Education*, 1(3), 175–198, DOI: 10.1080/15595690701394782

Russell, G. (2016) A variation on forced migration: Wilhelm Peters (Prussia via Britain to Turkey) and Muzafer Sherif (Turkey to the United States). *Journal of the History of the Neurosciences*, 25(3), 320–347, DOI: 10.1080/0964704X.2016.1175201

Şimşek, D. (2017). Göç politikaları ve 'insan güvenliği': Türkiye'deki Suriyeliler Örneği. *Toplum ve Bilim*, 140, 11–26.

Sözeri, E. K (2016). Two Petitions, Two Academia: Turkish Loneliness and the Universal Values. *Translate for Justice*, January. Retrieved from https://translateforjustice.com/2016/02/01/two-petitions-two-academia-turkish-loneliness-and-the-universal-values/

Taeuber, I. (1958). Population and Modernization in Turkey. *Population Index*, 24(2), 101–122. DOI:10.2307/2731516

Taştan, İ. Ö., Ördek, A., & Öz, F. (2020). *A Report on Academic Freedoms in Turkey in the Period of the State of Emergency*. Ankara: KAGED. Retrieved from https://www.ohchr.org/Documents/Issues/Opinion/Submissions/Academics/INSAN_HAKLARI_OKULU3.pdf

TİHV Akademi (2018). Barış İçin Akademisyenlere Yönelik Baskılar ve Hak İhlalleri. Retrieved from http://www.tihvakademi.org/wp-content/uploads/2019/03/BULTEN01.2018TR.pdf

Tekdemir, Ö., Toivanen, M., & Baser, B. (2018) Peace Profile: Academics for Peace in Turkey. *Peace Review, 30*(1), 103–111. DOI: 10.1080/10402659.2017.1419968

Tekeli, İ. (2019). Modern Türkiye'de bilim ve üniversite (1923-2019). In A. Şimşek (Ed.), *Modern Türkiye Tarihi*. (pp. 283–311). Pegem Yayınları.

Tekin, S. (2019). *Barış İçin Akademisyenler Vakasının Kısa Tarihi, 11 Ocak 2016–11 Ocak 2019*. Türkiye İnsan Hakları Vakfı Akademi. Retrieved from https://www.tihvakademi.org/wp-content/uploads/2019/03/Barisic inakademisyenlervakasi.pdf

The Science Academy (2017). The Science Academy Report on Academic Freedoms 2016–2017. *Bilim Akademisi*, August 9. Retrieved from www.bi limakademisi.org

Timur, T. (2000). *Toplumsal değişme ve üniversiteler*. İmge kitabevi.

Tomenendal, K., Özdemir, F. D., & Mercan, F. O. (2010). German-Speaking Academic Émigrés in Turkey of the 1940s. *Österreichische Zeitschrift für Geschichtswissenschaften, 21*(3), 69–99.

Tuğrul, B., & Deniz, E. (2019). Academies for Solidarity under the State of Exception in Turkey. In B. Tejerina, C. Miranda de Almeida & I. Perugorría (Eds.), *Sharing Society. The Impact of Collaborative Collective Actions in the Transformation of Contemporary Societies Conference Proceedings*. May 23–24, 2019. Universidad del País Vasco/Euskal Herriko Unibertsitatea, Bilbao, Spain.

Tutkal, S. (2020). Power, knowledge, and universities: Turkey's dismissed 'academics for peace'. *Critical Studies in Education*, 1–16.

Vatansever, A. (2018). Partners in crime: the anti-intellectual complicity between the state and the universities in Turkey. *The Journal of Interrupted Studies, 1*(1), 3–25.

Vatansever, A., & Yalçın, M. G. (2015). *Ne ders olsa veririz: Akademisyenin vasıfsız işçiye dönüşümü*. İletişim Yayınları.

Verşan, V. (1989). *Yüksek Öğretimde Değişmeler*. Ankara: Türk Eğitim Derneği Yayınları.

Yanardağ, A. (2017). Cumhuriyet dönemi darülfünun tartışmaları ve 1933 Darülfünun reformu. *Akademik Sosyal Araştırmalar Dergisi, 5*(6) 112–133.

Yapp, M. E., & Dewdney, J. C. (n.d.). Turkey. *Encyclopedia Britannica*. Retrieved from https://www.britannica.com/place/Turkey

Watenpaugh, K. D., Fricke, A. L., & King, J. R. (2014). *We will stop here and go no further: Syrian university students and scholars in Turkey*. New York: Institute of International Education, 284. Retrieved from http://www.sc

holarrescuefund.org/sites/default/files/pdf-articles/we-will-stop-here-a
nd-go-nofurther-syrian-university-students-and-scholars-in-turkey-00
2_1.pdf

Zürcher, E. J. (2000). *Modernleşen Türkiye'nin tarihi.* İletişim Yayınları.

Zürcher, E. J. (2017). *Turkey: A modern history.* London: I.B.Tauris.

The unheard voices
At-risk Syrian academics in Jordan, Lebanon, and Turkey

Nahed Ghazzoul

The Syrian refugee crisis has been one of the most significant exodus crises since the World War II (Ayvazoglu et al., 2021, p. 99). Worldwide, Syria occupies the first rank in the number of refugees (6.6 million) and internally displaced persons (5.6 million) (UNHCR, 2021). The global response to the surge of Syrian refugees has focused on addressing their most immediate needs, such as shelter, food, primary medical care, and primary education, whereas other critical issues, such as higher education and the situation of displaced Syrian academics, remain understudied. Very little is known about the personal experiences and professional trajectories of scholars who left Syria following the start of the conflict in 2011. This is particularly true for those scholars who resettled in neighboring countries, such as Turkey, Jordan, and Lebanon.

Tigau (2019) has related the lack of research in this field to the difficulty of collecting data in times of conflict and displacement. Another reason has been attributed to the small number of displaced academics in comparison with the overall number of displaced and refugee groups. Hence, the main attention has been devoted to the education of the young generation from primary to secondary schools (Štrajn, 2014), leaving the situation of displaced scholars unaddressed (Salehyan, 2019, p. 146).

This chapter intends to narrow this gap by providing a cross-country comparison of the situation and professional pathways of displaced Syrian academics in Jordan, Lebanon, and Turkey. It is based on an original survey comprising responses of Syrian academics living in the three countries. The survey data are complemented by personal encounters and observations of the author, who has lived and worked in exile after leaving Syria.

The following sections provide a review of the forced displacement of Syrian academics, present the conducted survey, and contextualize the findings. As a point of departure, academics are defined as PhD and MA degree holders who, before leaving Syria, were instructors at public or private universities and colleges accredited by the Syrian Ministry of Higher Education. The chapter concludes by offering recommendations to international support organizations for at-risk and displaced scholars.

Forced Displacement of Syrian Academics

With the outbreak of the conflict in Syria in 2011, disruption started to greatly impact the entire educational landscape. Syrian academics were targeted by most of the armed parties to the conflict. The universities suffered from "politicisation, militarisation and human rights violations" (University World News, 2019, p. 1). Academics were assaulted, threatened, kidnapped, assassinated, or detained by the regime (Matthews, 2013). Confirming this, Pherali and Millican (2020) reported that "a large number of academics have been either exiled or killed; any research that existed has almost disappeared and teaching has been disrupted by absenteeism, lack of resources and limited numbers of experienced faculty members" (p. 6).

Among those who lost their lives during the early years of the conflict were Dr. Mohammad Alomar (October 2, 2011); Dr. A'ws Abdul Karim Khalil, nuclear physicist (September 2011); Brigadier General Dr. Nael Al-Dakhil, academic director of the Military Chemistry College (September 2011); Dr. Eng. Muhammad Ali Aqil, professor at the Faculty of Architecture at Al-Baath University (September 2011); Dr. Mahmood Tasabihji in 2012 at the Faculty of Medicine; Dr. Colonel Suheel Shahin and his colleagues at Al-Assad Academy for Military Engineering (September 2012); Dr. Ilias Issac, professor of agricultural engineering (September 2012); and Dr. Khaled al-Asaad, scholar of antiquities and Aramaic culture (August 2015), to mention only a few.

Assassinations of academics were reported in Aleppo and Homs (Matthews, 2013). Some scholars were beaten and dragged off to prison by security forces in front of their students (McLaughlin et al., 2020). According to Matthews (2013), Syrian universities were used as detention centers, while both staff and students were targeted. Freedom of expression or academic criticism of the political, security, and social situation was forbidden. Academics and students alike were frequently reported by their colleagues, students, or intelligence members who were "planted" in every lecture room. As a result, many academics were dismissed from work and imprisoned or went missing (Masud, 2018). At the same time, students and scholars staying abroad were stranded, had dwindling funds, and could not return to Syria.

The conflict has led to massive losses of both research expertise and higher education infrastructure (University World News, 2019). A substantial number of Syrian academics have left the country. There are no concrete figures or accurate numbers, but according to CARA, an estimated number of 2,000 Syrian academics fled the conflict when it began (Beall & Lonsdale, 2019, p. 5). Confirming this, Al-Ibrahim (2016) stated that the number of Syrian academics employed at the five public universities was about 7,500 professors, and the number of those who left the country as of 2016 was about 2,000. He added that around 500 professors were internally displaced in the liberated areas. Furthermore, hundreds of PhD students at Syrian public universities, who were appointed as full-time instructors, mostly went abroad or were also internally displaced (Al-Ibrahim, 2016).

An official at the Ministry of Higher Education stated that more than 60% of Syrian academics had left Syria by the end of 2017. This has led to a significant shortage in faculty members, making departments function at a capacity of less than 40% (Sputnik, 2017). Echoing this, the president of Aleppo University stated that more than two-thirds of the faculty members had left the university (Sputnik, 2017).

In response to the massive exodus of scholars and intellectuals and to prevent further emigration, in 2014, the Syrian regime suggested considering academics who left the country (irrespective of the reason) as having resigned, and promised to put them on trial once they returned. In a meeting held at Damascus University in 2014, the president of the university called academics who fled the conflict traitors of their country (Enib Baladi, 2014). This, however, has only added to the insecurities experienced by scholars, and the large-scale emigration of Syrian academics continued, leading to the fragmentation of the Syrian higher education system and its near collapse (Al Hessan et al., 2016; Masud, 2018).

Despite the fact that there is an increased global awareness of the importance of higher education, Syrian academics receive only limited support from the international community (Parkinson et al., 2020). The most prominent academic institutions and networks that have offered help to

Syrian academics include the Institute of International Education's Scholar Rescue Fund (IIE-SRF), an international Scholars at Risk network (SAR), the British NGO CARA, the French Programme national d'aide à l'Accueil en Urgence des Scientifiques en Exil (Programme for the Emergency Reception of Scientists in Exile, PAUSE), and, recently, the Mellon Foundation through funding a fellowship for Syrian academics at the Columbia Global Centre in Amman. Yet, since the scale of the tragedy is so immense, the support offered to Syrian academics is far below what is needed.

Starting a new life in exile brings new challenges, which Syrian academics had not necessarily anticipated before they left, such as the reality of being (perceived as) refugees. Having escaped the war and resettled in other countries, many Syrian academics do not accept being identified as refugees because it strips them of practicing their agency and reminds them of "being rootless, stateless and rejected" (CARA, 2019, p.78). In their eyes, the international system labels refugees with this title to comfort the host country citizens and to indicate that the newcomers are less competitive than locals, and that they are vulnerable and need assistance. Thus, the relationship of those receiving the refugee label in host communities is not based on partnership, collaboration or shared experiences; rather, it is based on the feeling of inferiority. As one Syrian academic, interviewed by Pherali (2020, p. 94), put it:

> [A]refugee is a downstream labeling. When you get this label, you automatically become inferior. Wherever you are, whatever you do. You are second to the counterpart. It does not matter being an academician because being a refugee already makes you a less worthy person.

The refugee label violates all other social and professional identities an academic has. It is a dismissive term that does not embody their self-respect and dignity. Therefore, many displaced Syrian academics try to avoid registering with the UN refugee agency, the UNHCR, and receiving a refugee ID card, or if this is unavoidable, they keep it secret, so that they are not seen as inferior to other academic colleagues.

Mass displacement of Syrian academics, scientists, and intellectuals has exacted heavy tolls on Syria, not only in terms of the loss of human capital, but also on the destruction of the intellectual life of the Syrian community. At the same time, in many cases, it has contributed to the intellectual and scientific life of the host societies, and to the academic transformation and internationalization in the academic environment of the host institutions

(Elsner, 2017). However, as this study found, this was not the case for Syrian academics, as they were often unable to continue their careers due to legal restrictions, economic conditions, language barriers, psychological trauma, or unrecognized qualifications (Parkinson et al., 2020, p. 185).

The survey

This study aimed to generate a better understanding of the current professional situation of Syrian displaced academics living in Jordan, Lebanon, and Turkey. To that end, an original survey was developed for and conducted among Syrian academics in the three countries, asking about their experiences in exile. The questions focused on various aspects of life, including their employment status, and whether they have benefited from any professional development programs or research opportunities in the host countries. Most importantly, the survey concentrated on the following issues: work-related challenges experienced by displaced Syrian academics in Jordan, Lebanon, and Turkey, their employment prospects, and support programs for displaced scholars.

The total number of respondents comprised 126 academics, including 113 males and 13 females. As for the distribution of qualifications, 90 participants were PhD holders and 36 master's degree holders. In terms of academic rank, 19 academics were full professors, 20 associate professors, 44 assistant professors, and 43 lecturers. As for host countries, 71 respondents resided in Turkey, 38 in Jordan, and 17 in Lebanon.

Notably, the number of female respondents was significantly smaller than the number of male respondents, which reflected the respective gender gap existing in academic faculty employment at Syrian universities before 2011. This gap was the result of cultural factors which do not encourage higher education of females, and the Syrian Education Ministry's policy of promoting education and career development for men, which created an extremely unfair competitive environment for female academics versus their male colleagues.

Responding to the survey questions, 70.6% of the participants indicated that they did not have a job in their country of exile. However, the current employment varied considerably according to their academic rank. For example, 12 full-professors out of 19 (63%) mentioned that they were employed, and only two associate professors out of 20 (10%) indicated that they had a job. As for assistant professors 34 out of 44 (77%) do not work, and 35 out of 43 (81%) lecturers revealed that they do not work.

Overall, full professors (males) had considerably better chances of finding employment in their countries of exile. The data further revealed that out of 90 PhD holders, 61 were not employed whereas only six out of 34 master's degree holders declared that they worked, which in neither case did not necessarily mean that they worked in the academic sector. When asking the participants if they had benefited from any professional development program since they left Syria, 67% responded positively, whereas only 11% obtained fellowships or support for a research stay.

When asking the participants about the reasons for not being employed, the common answers from Jordan, Lebanon, and Turkey were the lack of opportunities and having a refugee status, which implied legal constraints in the job market. The latter was a particular challenge in Lebanon and Jordan, where Syrian refugees are largely denied access to white-collar jobs. As for participants based in Turkey, the challenges were at times related to the government's policies, which one would have expected to benefit displaced scholars. Thus, the Turkish government granted citizenship to some Syrian academics and highly qualified Syrian intellectuals, which one would assume could open new doors for academic career development. However, this had an opposite effect on the scholars' work status. Suddenly, they were treated as local academics, and their work contracts previously signed under the status of displaced scholars were terminated immediately after receiving Turkish nationality. They then had to be enrolled on the job waiting lists for the local faculty. Another challenge was related to the language barrier in both Turkey and Lebanon. Turkish and English are the languages of instruction in almost all disciplines at Turkish universities, a skill many Syrian academics lack. In Lebanon, academics were required to teach in English and French in as well as Arabic. So, in addition to political, economic, and security concerns, language differences became a serious obstacle.

When answering the question concerning their future career prospects, all the participants responded that the current lack of opportunities affected them negatively. Some stated that, with the lack of current employment, their academic qualifications were deteriorating and their knowledge was far from being updated, and two respondents, in a fit of despair, questioned why they had studied and got their PhD in the first place. Even those who were employed in the higher education sector saw a limited potential for further development of their academic careers in host environments that were rarely, if ever, supportive of their academic promotion.

Finally, when asked to offer suggestions to international support organizations for at-risk and displaced scholars, most of the respondents emphasized the need for assistance with creating job opportunities, fellowships, access to libraries, collaborative research partnerships with foreign academics, capacity building, and intensive English and Turkish language courses. In answering the final question about adding extra information, they all called for urgent and immediate help. They felt that they had been forgotten by the international community.

Contextualizing the survey findings

Overall, the situation of Syrian academics in exile was not what they had expected when they left Syria. Their qualifications were not valued in the job market. They were often perceived as a threat to local academics who, at times, acted in a quite hostile manner, for instance, in Jordan by publicly requesting to "Jordanize" the higher education sector. The challenges faced by Syrian academics in exile ranged from political, social, and financial issues to legal constraints pertaining to their refugee status. Contextualizing the above survey findings, this section provides a broader picture of the differences across the three host countries and the respective support programs in place.

Syrian academics in Jordan

Prior to 2011, Jordan was the biggest "importer" of Syrian scholars. Syrian academics were paid generously, and many inducements and facilities were offered to them, including the possibility of teaching for two days a week and returning to Syria. They were perceived as rare experts, and the university that attracted the largest number of Syrian academics would attract the biggest number of students. Syrian academics were considered competent intellectuals who participated in building the Jordanian higher educational system.

However, after 2011, this situation gradually changed. This could have been related to various factors, including the fact that hundreds of Syrian academics had arrived in Jordan, which made the supply in the job market much bigger than the demand. Jordanian academics started considering

Syrian counterparts as a threat, not as an added value. Consequently, calls for Jordanizing the higher education system increased. The situation worsened after some Gulf countries' universities terminated contracts with many Jordanian academics due to economic factors after 2016.

Lack of job opportunities and financial support, soaring prices, and the inability of their children to enroll in public schools from 2013–2015 led a significant number of Syrian academics to leave Jordan. They moved to Turkey, Europe, and the Gulf countries when visas were still available for Syrians. Thus, the number of Syrian academics at Jordanian universities has gradually dropped, along with their salaries, starting from the 2016 academic year onwards. According to official information released by the Jordanian Accreditation Corporation, the number of Syrian academics who worked at Jordanian universities from 2017–2021 has decreased gradually from 63 in 2017–2018 to 20 in 2020–2021.

Furthermore, the work permit fees for Syrian academics increased in 2019 from $10 to $3,600 based on Decree Number 2019/290 (Jordanian Ministry of Labor, 2019). This financial burden can be seen as a message to Jordanian universities to end hiring of non-Jordanian academics. Most Jordanian universities required the Syrian academics to pay for their work permit, or to share half the payment with the university, in addition to the late payment fine. It is worth mentioning that the Syrian academics who sustained their jobs were either full professors or associate professors, whom the university could not replace with a Jordanian academic in the same specialty or the same academic rank. In 2021, this number has decreased as more Syrian academics 'contracts have been terminated. Currently, displaced Syrian academics have even fewer job opportunities in the Jordanian academy, and thus, support by international organizations will be critical in assisting them in finding positions (Watenpaugh et al., 2013, p. 6).

Syrian academics in Lebanon

Prior to 2011, Syrian academics were able to work as part-time faculty members at Lebanese universities, getting all the support needed to obtain work permits and sometimes even being able to work without them. Many taught at Syrian and Lebanese universities simultaneously. However, things changed dramatically after 2011 and Syrian academics found it difficult to get jobs and faced social, political, and legal obstacles. According to Reisz (2014), the situation of Syrian academics and students in Lebanon is generally bleak.

They face "severe resource constraints and physical threats, [and] unwritten discriminatory policies make Syrian students and academics vulnerable to exclusion from higher education in Lebanon due to unwritten discriminatory laws" (Watenpaugh et al., 2014a, p. 13).

When trying to complete their residency permit in Lebanon, Syrian academics must sign a pledge at a public notary stating that they will not work or apply for work in Lebanon, as noted by the academics who participated in the study. Therefore, the legal impediments to employment play a decisive role in finding jobs in Lebanon. The process of obtaining work permits is burdensome and very seldom successful (Bidinger et al., 2015, p. 45). The presence of Syrian military troops in Lebanon for a period of 29 years, from the beginning of 1976 until April 2005, and the indirect control of the Syrian regime of all decision-making processes to date have created hostility toward hiring Syrian academics in Lebanon. In addition, as suggested by Watenpaugh et al. (2014a), the primary challenge Syrian scholars face in the Lebanese job market is "exacerbated by their generally weaker English and French proficiency compared to their Lebanese counterparts" (p. 26).

Furthermore, the complex Lebanese political situation dramatically affects the labor market. Syrian scholars and academics, who oppose the Assad regime, fear the ability of the Syrian intelligence to reach the Lebanese universities to capture them (Reisz, 2014). Administrators at several Lebanese academic institutions are afraid to hire any Syrian academic who possibly opposes the Assad regime, either because they support the regime or because they fear political trouble. If an academic is hired, they must sign a pledge that they will not be involved in any political activity inside or outside the university, as reported by most of the Syrian academics who participated in the survey for this study. They also have to agree not to discuss any political issue during the courses they teach, which clearly violates principles of academic freedom (Watenpaugh et al., 2014a).

Syrian academics in Turkey

According to the UNHCR (2020), Turkey has accepted the largest number of Syrian refugees worldwide, more than 3.6 million, including the largest number of Syrian academics. The Turkish government estimated that 1000 Syrian academics resided in the country in 2016; this included only PhD holders (İçduygu & Millet, 2016).

Despite the fact that the Turkish government allowed the appointment of Syrian academics in Turkish universities, the challenges and restrictions they faced were enormous. For example, like other Syrian refugees, Syrian academics were not allowed to travel outside Turkey. Also, they were prevented from taking certain paid positions due to their temporary refugee status (İçduygu & Millet, 2016), or because they could not teach in Turkish or English languages, or they lacked the necessary official personal documentation (Ammar, 2016).

Other challenges included academic equalization of their master and PhD degrees, which at times took years. Also, in cases where they did not have an official copy of their certificates because they had fled the war without bringing these, the Turkish Ministry of Higher Education could not hire them in Turkish higher education institutions, or issue work permits to them. In addition, the decentralized nature of Turkish higher education, a situation exacerbated by language barriers as Turkish is the main language of instruction (Watenpaugh et al., 2014b, p. 5), deprived many Syrian academics of the opportunity to teach at Turkish universities. Due also to a lack of language proficiency, they could not work in international study programs where English was the main language of instruction. However, academics whose specialties were Arabic language or Islamic or theology studies were offered job opportunities because the language of instruction was Arabic. Currently, requests to establish universities that teach in Arabic to accommodate displaced Syrian students, and provide teaching opportunities for Syrian academics have increased. Moreover, the Turkish government established departments at seven universities, where the language of instruction is Arabic. Additionally, Gaziantip University started BA/BSC and postgraduate programs in Arabic in 2015 (see also Hünler in this volume).

International support

Responding to the needs of the Syrian academics in Turkey, CARA's Syria Programme (CARASP) focused on three strategic components: English for academic purposes, research incubation [visits], and academic skills development (ASD) (Parkinson, 2018, p. 7). The first two were clear, but very little information was known about the type of academic development the Syrian academic participants needed to receive from CARASP. Therefore, CARA started gathering information through surveys and interviews to inform subsequent actions. It mobilized financial support and provided English language courses, in addition to some programs that aimed to develop their proficiency, such as "CARA Syria Programme E-Learn Soiree Sessions" which continue to the date of publication of this text. These programs provided Syrian academics and highly qualified Syrians with training in research methods curriculum design, and offered them "academic space where Syrian colleagues were able to disseminate their work and continue engaging in academic activities" (Pherali & Millican, 2020, p. 4). In addition, they facilitated channels of academic collaboration with colleagues in the UK, who have supported CARASP, and Syrian academics in authoring joint research papers, or reviewing their papers and helping in the publication process.

Despite this support, many Syrian academics in exile are facing dire circumstances and considerable challenges to continue their career; this includes psychological collective traumas, visa issues, accreditation problems, lack of financial resources, and isolation from scientific communities (Parkinson, 2018, p. 7). Only individual Syrian academics currently living in Turkey and some of those living in Jordan can benefit from such programs as those offered by CARA. Overall, the findings show that the academics based in Turkey received the most support in terms of professional development programs and English classes. In contrast, support for Syrian academics living in Lebanon and Jordan has been extremely limited. For instance, in 2020, the Mellon Foundation offered a 12-month fellowship program for displaced scholars, in which only one or two Syrian academics based in Jordan could join. Many could not apply for it due to the English language requirement and the focus of the fellowship on social sciences. In general, because most support programs and fellowships for at-risk scholars are based on English, this dramatically limits the number of Syrian applicants.

Furthermore, acceptance criteria for such support programs are often difficult for Syrian scholars to meet. Thus, they are only eligible to apply for the IIE-SRF program during the first three years after leaving their home country. Moreover, many Syrian academics who moved to neighboring countries learned about the existence of these programs too late, months or even years after arriving in Jordan, Lebanon or Turkey. In addition, these criteria do not take into account that displaced scholars need time to settle in with their families, go through security procedures in host communities, look for schools and jobs in different academic institutes, try to equalize their credentials, and look for official personal identification cards, so time passes, and the three-year condition for applying for IIE-SRF is not met.

Compared with refugee academics based in Northern Europe, who might be considered fortunate given their personal agency, extended cultural and social capital, and access to basic services, as well as a greater range of support networks (Pherali, 2020, p. 88), the situations for refugee academics in Jordan, Lebanon, and Turkey are completely different. In these countries, displaced Syrian academics often struggle to survive. Those who are registered at the UNHCR are classified as less vulnerable, so they are often exempted from much of the support provided by this organization. However, they share with the rest of refugees limited job access and spatial mobility and as noted by Pherali (2020, p. 88), "political freedom and uncertainty around their futures are similar to general refugee populations who are stuck in camps or host communities."

Conclusion

This chapter has discussed the difficult situation of displaced Syrian academics in Turkey, Jordan, and Lebanon. The findings of the study indicate that there is a lack of support for Syrian academics in the three countries; however, the Syrian academics in Turkey are in a somewhat better position than the others. They also show that each country needs more direct international responses, according to the country-specific situation. However, national and international responses should be discussed with the target communities, namely, the displaced Syrian scholars themselves, to reach the optimal goals.

It is worth mentioning that humanitarian organizations, such as the Norwegian Refugee Council, have conducted studies about the livelihood of Syrian refugees. Such studies have been conducted with the help of academics in the host countries; however, Syrian academics are rarely consulted or involved in preparing such researches. For example, several studies have been conducted in Jordan and Lebanon to explore the educational needs of the Syrian refugees, or their livelihood, but none of these studies engaged Syrian researchers. They always depend on other nationalities to speak on behalf of the Syrians rather than asking Syrians to play an active role in the studies. Engaging Syrian academics in these educational and research activities would help advance their careers, provide them with a source of living even if temporarily, and provide the field of studies with more accurate and authentic data relevant to the Syrian situation.

Support networks and programs such as PAUSE, CARA, SAR, and SRF, as well as universities worldwide, could play more critical roles by sponsoring job opportunities, online teaching, fellowships, research projects, and partnership programs with host universities in Turkey, Jordan, and Lebanon. Providing access to libraries and databases, and offering professional development programs, in addition to English for academic purposes courses, would be of great benefit for displaced Syrian academics.

Notably, most countries refuse to grant entry visas to Syrian academics to attend conferences or join postdoctoral fellowships. Such actions diminish their international academic values and prospects for knowledge sharing and exchange. Promoting international research collaborations would help Syrian academics overcome the isolation of exile and bridge potential skill and knowledge gaps. Engaging them in capacity-building programs in host countries would ultimately help them to rebuild the higher educational system in Syria once they have the opportunity to return.

References

Al Hessan, M., Bengtsson, S., & Kohlenberger, J. (2016). Understanding the Syrian educational system in a context of crisis. Vienna Institute of Demography Working Papers No. 09/2016) Austrian Academy of Science (ÖAW), Vienna Institute of Demography (VID). Retrieved from http://hdl .handle.net/10419/156317

Al-Ibrahim, A. (2016, February 28). Syrian Professor's Plea: Make Us Part of the Solution. *Al-Fanar Media*. Retrieved from https://www.al-fanarmedia. org/2016/02/syrian-professors-plea-make-us-part-of-the-solution/

Ammar, J. (2016). Interview with Jamil Ammar. *International Journal of Research from the Front-Line*, 1(2), 156–162.

Ayvazoglu, A.S., Kunuroglu, F., & Yagmur, K. (2021). Psychological and Socio-Cultural Adaptation of Syrian Refugees in Turkey. *International Journal of Intercultural Relations*, 80, 99–111.

Beall, J., & Lonsdale A. (2019). *The State of Higher Education in Syria Pre-2011*. Cambridge: Cambridge University Press.

Bidinger, S., Lang, A. Hites, D., Kuzmova, Y., Noureddine, N., & Akram, S.A. (2015). *Protecting Syrian Refugees: Laws, Policies, and Global Responsibility Sharing*. Boston: Boston University School of Law.

Council for At-Risk Academics – CARA (2019). Syrian higher education post 2011: Immediate and future challenges. Retrieved from https://www.cara .ngo/wp-content/uploads/2019/06/190606-REPORT-2-POST-2011-FINAL -ENGLISH.pdf

Elsner, J. (2017). Pfeiffer, Fraenkel, and refugee scholarship in Oxford during and after the Second World War. In S. Crawford, K. Ulmschneider & J. Elsner (Eds.), *Ark of civilization: Refugee scholars and Oxford University, 1930–1945* (pp. 26–49). Oxford: University of Oxford.

Enib Baladi (2014). Damascus University: The Decision to Impose Fines on Violators is not New (in Arabic): Retrieved from https://www.enabbaladi. net/archives/351474.

İçduygu, A., & Millet, E. (2016). *Syrian Refugees in Turkey: Insecure Lives in an Environment of Pseudo-Integration*. Istanbul: Koç University Press.

Jordanian Ministry of Labor (2019). Website. Retrieved from http://www.mol .gov.jo/AR/Pages/

Masud, M. (2018). Authoritarian Claims to Legitimacy: Syria's Education under the Regime of Bashar al-Assad. *Mediterranean Studies, 26*(1), 80–111.

Matthews, D. (2013). Calls for Urgent Support for Syrian Academics. *Times Higher Education*. Retrieved from https://www.timeshighereducation.com /cara-calls-for-urgent-support-for-syrian-academics/422280.article

McLaughlin, C., Dillabough, J., Fimyar, O., & Al Azmeh, Z. (2020). Testimonies of Syrian Academics Displaced Post 2011: Time, place and the agentic self. *International Journal of Educational Research Open*, 1, https://doi.org/10.1016/ j.ijedro.2020.100003

Parkinson, T. (2018). A Trialectic Framework for Large Group Processes in Educational Action Research: The Case of Academic Development for Syrian Academics in Exile. *Educational Action Research, 27*(5), 798–814.

Parkinson, T., McDonald, MC., & Quinlan, K. M. (2020). Reconceptualising academic Development as Community Development: Lessons from Working with Syrian Academics in Exile. *Higher Education, 79*, 183–201.

Pherali, T. (2020). My life as a second-class human being: Experiences of a refugee academic. *Education and Conflict Review, 3*, 87–97.

Pherali, T., & Millican, J. (2020). Rebuilding Syrian higher education for a stable future. *Education and Conflict Review, 3*. https://www.cara.ngo/wp-c ontent/uploads/2020/08/Conflict_Review_2020_final_web.pdf

Reisz, M. (2014, June 26). Refugees from Syria Excluded from Lebanese Universities.*Times Higher Education.* Retrieved from https://www.timeshi ghereducation.com/cn/news/refugees-from-syria-excluded-from-lebane se-universities/2014142.article

Salehyan, I. (2019). Conclusion: What academia can contribute to refugee policy. *Journal of Peace Research*, 56(1), 146–151.

Sputnik (2017). 20% of the Syrian Academics have left Syria and some would like to return back (in Arabic). *Sputnik News.* Retrieved from https://arabi c.sputniknews.com/arab_world/201705101023935680

Štrajn, D. (2014). Book Review of: Bacevic, Jana, From class to identity: The Politics of education reforms in former Yugoslavia. *CEPS Journal*, 4(4), 163–166.

Tigau, C. (2019). Conflict-induced displacement of skilled refugees: A cross-case analysis of Syrian professionals in selected OECD countries. *Norteamérica, Revista Académica del CISAN-UNAM*, 14(1), 341–368.

United Nations High Commissioner for Refugees – UNHCR (2020). Global Trends Forced Displacement in 2019 Report. Geneva. Retrieved from http s://www.unhcr.org/5ee200e37.pdf

United Nations High Commissioner for Refugees – UNHCR (2021). Syria Emergency. Retrieved from https://www.unhcr.org/syria-emergency.ht ml

University World News (2019, June 19). HE system left broken by conflict, academics targeted. *University World News.* Retrieved from https://www. universityworldnews.com/post.php?story=20190619074049474

Watenpaugh, K., Fricke, A., King, J., Gratien, C., & Siegel, T. (2013). *Uncounted and Unacknowledged: Syria's Refugee University Students and Academics in Jordan.* Davis: University of California Davis Institute of International Education.

Watenpaugh K. D., Fricke, A. L., King, J., Arrar, R., Siegel, T., & Stanton, A. (2014a). *The War Follows Them: Syrian University Students and Scholars in Lebanon.* New York: Institute of International Education.

Watenpaugh, K.D, Fricke, A., King, J., Gratien, C., & Yilmaz, S. (2014b). *We will stop and go no further: Syrian University Students and Scholars in Turkey.* University of California Davis: Institute of International Education.

Academic freedom and the untold story of Venezuelan scholars under pressure

David Gómez Gamboa & Lizzy Anjel-van Dijk

Since 2010, Venezuela has faced a severe humanitarian crisis and growing restrictions on civic freedoms, forcing an estimated 5–6 million Venezuelans to leave the country. This accounts for 15%–20% of the country's population currently living abroad. A significant number of the people who see themselves as forced to migrate are highly qualified academics with postgraduate degrees, including postdoctoral researchers and university professors. Their situation and experiences of living in exile, however, rarely find their way into scholarly analyses. In this chapter, we address this gap by examining the factors that force Venezuelan scholars to flee their home country as well as the opportunities and challenges they encounter in their host countries. We focus specifically on the exodus of Venezuelan scholars during 2014–2020, the period when the emigration of Venezuelans increased dramatically due to the worsening political and humanitarian crises as well as the widespread violations of human rights.

Methodologically, the research draws on a systematic review of secondary literature and country reports by organizations such as Aula Abierta and Scholars at Risk (SAR) as well as insights from original semi-structured interviews conducted in online meetings with Venezuelan scholars in exile. The interviews were conducted from August through November 2020 using a snowball sampling technique identifying 230 Venezuelan scholars living abroad. Of this group, 50 scholars were approached for semi-structured interviews to gain a more in-depth understanding of their experiences of living and working in exile. Here, we aimed for a sample of Venezuelan professors from different universities and from a variety of disciplines who currently reside in one of the three main destination countries for highly qualified Venezuelans: Colombia, Ecuador, and Chile. All quoted interviewee statements in this chapter are translated from Spanish to English.

In the following sections, we first discuss the current situation of an unprecedented exodus of Venezuelan academics, which is followed by an analysis of the push and pull factors that cause them to migrate. We then synthesize the exile experiences of the scholars we interviewed to reveal the challenges they face in their host countries. We conclude with the implications of Venezuelan scholars migrating, including opportunities for the internationalization of knowledge as well as for reviving national academia when the return of scholars to Venezuela becomes possible.

An unprecedented exodus of Venezuelan scholars

Three different migration flows can be identified throughout last decades out of Venezuela. The first flow occurred from 2007 to 2013 after the reelection of president Hugo Chávez who pushed for constitutional reform. This period has been described by several academics who were interviewed as the "beginning of a new phase of authoritarianism in Venezuela" (personal interviews, August 24–30, 2020). Due to increasing chaos, Colombo-Venezuelan scholars in particular left Venezuela and migrated to Colombia. Colombo-Venezuelans have both nationalities as many Colombians who had to flee to Venezuela in earlier years due to the internal conflict in Colombia. The second migration flow followed from 2014 to 2019. This was caused by serious human rights violations during the protests of 2014 and 2017 and the worsening of the humanitarian emergency resulting from a severe shortage of food and medicines.[1] Besides the scholars with a dual nationality, more scholars saw themselves as being forced to leave the country due to human rights violations and the dire humanitarian situation. Finally, the third wave occurred from 2019 to 2020. The reasons given for this during the interviews were the ever worsening humanitarian situation, including the collapse of electricity systems and public services across the country (Aula Abierta Venezuela, 2020c, 2020d).

While the socioeconomic situation heavily affects the education and research system in Venezuela, neither the Education Ministry nor Higher

---

1   For several years the Inter-American Commission on Human Rights (IACHR) has observed a gradual deterioration in the democratic institutional system and the human rights situation in Venezuela that has significantly intensified and become widespread since 2015 (IACHR, 2017a, 2017b; OAS, 2018).

Education Ministry has presented credible annual accountability reports as of 2016. This bespeaks a public policy of intentionally not informing and disinforming the public, including cases of disseminating unsupported information and even information that is proven to be contradictory (Gómez Gamboa et al., 2020a; Programa Venezolano de Educación Acción en Derechos Humanos, 2017). Additionally, there is no reliable information about how many Venezuelan scholars have emigrated. A statistical study by Requena and Caputo (2016)[2] determined that 13% of the total researchers in the country have been forced to leave their jobs. It is estimated, however, that the numbers are much higher (personal interviews, August 24-30, 2020).

From 2017 to 2020, alarming levels of professor dropout rates from 30%–50% were reported by Aula Abierta (2019a), a non-governmental organization (NGO) that works for the promotion and defense of human rights in the education sector within Latin America. Resignation by and requests for sabbaticals from professors at the universities became recurrent. In 2019, the Federation of Associations of University Professors of Venezuela (FAPUV) placed the dropout rate of university professors at 40% since 2013. The SAR's Free to Think Report of 2020 stated that by 2019, approximately 50% of professors from all Venezuelan universities had left the country. Likewise, 30% of the country's researchers (who work in labs or research centers) had emigrated by April 2019 (El Nacional, 2019). By 2018, budget constraints and the migration of researchers had reportedly left 77% of laboratories in Venezuela—including labs that until recently were major contributors to the country's public health system—paralyzed or abandoned (Conte de San Blas, 2018; SAR, 2020; University of Los Andes Human Rights Observatory, 2020). This represents dire circumstances, especially amid the spread of COVID-19.

To indicate the severity of the situation, from 2017 to 2020, the Law School of the University of Zulia (LUZ), one of the major universities in the country, suffered a 50% decrease in its faculty. Of 134 active professors in 2017, only 68 professors remained active in 2020. Counting the professors who resigned (18) and those who requested permits and sabbaticals (11), the number of

---

2   According to Requena and Caputo (2016), 1,670 Venezuelan scientists (from 1960–2014) have been forced to leave the country. They constituted 13% of the total research publication community, which was made up of 12,850 research publishers until 2014. They were responsible for the production of 13,240 accredited publications, or 28% of the national total of such publications produced by the country from 1960 to 2014 (42,783 publications).

professors who left the Law School of LUZ was 29 (21.6%) during this period, of which the majority stated that they intended to migrate to another country. The rest comprised 30 (22.3%) professors who retired during 2017–2020, and seven (5.2%) who passed away.

Simon Bolívar University, with one of the most prominent engineering programs in the country, had only 26 research programs left in 2018; in 1997, it had conducted 165 research programs. According to the National Observatory of Science and Technology, in 2012, the university financed 974 research projects, whereas only 62 projects were financed in 2015 (SAR, 2020, p. 111; Red Iberoamericana de Indicadores de Ciencia y Tecnología, 2018).

The emigration of doctors (professors or students from postgraduate medical programs) generates serious risks to public access to health services in the country. Along with the decrease in medical specialists in Venezuela, graduate students in 2019 also showed an increased tendency to migrate after completing their studies. At least 50% of graduates in 2016 emigrated in 2017. One hundred percent of the graduates from the specialty of otorhinolaryngology at University of Zulia (LUZ) migrated, 80% of graduates of thoracic surgery migrated, and 75% from the program on ICU of the LUZ medicine postgraduate program left the country as well.[3] Notably, the brain drain of medics in the context of the current pandemic has had devastating effects on the public health situation in the country.

The massive exodus of scholars has also had a drastic effect on the overall research production of Venezuelan universities and their position in international university rankings. From 2017 to 2020, this effect was most obvious: based on the QS Latin American University Rankings, Venezuela's position fell from number 115 to number 160 (QS, 2020a, 2020b; Scimago Institutions Rankings, n.d.-a). According to the SAR's Free to Think Report (2020), the country's research output (the number of scientific publications) has also declined significantly since the start of the 2000s. Cited in the SAR's report, SCImago Journal & Country Rank, a publicly available online portal that ranks countries by journal output, revealed that Venezuelan universities provided 4.8% of the journal articles from Latin America in

---

3   During 2019 and 2020, an alarming dropout level devastated the postgraduate medicine programs at LUZ in cardiovascular surgery (71.4% dropout rate). The average number of students of every postgraduate course at LUZ is not even four students, while every program is designed for 25 students. That is 84% fewer students than were expected to sign up.

1998 (roughly proportional to Venezuela's population as a percentage of the overall population of Latin America at the time) (Worldometers, n.d.). By 2019, that number dropped to 0.8% (Scimago Institutions Rankings, n.d.-b). Thus, if in the late 1990s Venezuela's higher education and research system occupied one of the highest positions in Latin America, the severe humanitarian crisis, political repression, and resultant massive exodus of Venezuelan academics throughout the 2010s have left the country with few prospects for the development of science and progressive reforms in higher education.

Push factors for the exodus of Venezuelan scholars

The United Nations High Commissioner for Refugees (UNHCR) has identified the lack of democracy and the severe humanitarian crisis as the main causes of the unprecedented exodus of Venezuelans.[4] The situation in the Venezuelan higher education sector is indicative of these trends. After Hugo Chavez came to power in 1999, the higher education and research system became an ideological instrument in the hands of the country's leadership, and critical voices at universities were systematically targeted by the "revolutionary government" via both economic and political means, ultimately driving many intellectuals out of the country. In addition, the entire education sector faced a gradual reduction in faculty salaries and social benefits. Autonomous public universities had their operational budgets frozen, and staff salaries were not paid despite enormous hyperinflation.[5] If prior to the 2000s, getting a university degree and working in the higher education sector was one of the main pathways for upward social mobility, by 2010 university employees faced an unprecedented decline in their economic resources and social prestige.

The situation worsened considerably in 2016 with the onset of the humanitarian crisis. From 2017 to 2020, a monthly university professor's salary in Venezuela ranged from US$3 to US$20 (Comisión Interamericana

---

4   The UNHCR has identified it as "One of the largest population exodus in The Americas since 1950" (Bastidas, 2018). The United Nations High Commissioner for Human Rights on March 7, 2018 stated that he was deeply disturbed by the growing exodus of Venezuelans (IACHR, 2018; VenEuropa Canal, 2018).

5   For 2019, hyperinflation in Venezuela was around 10 million percent according to the International Monetary Fund (Hanke, 2019).

de Derechos Humanos, 2019; IACHR, 2019, p. 197), which was barely enough to survive. In September 2018, the National Assembly declared a humanitarian emergency in the education sector (Barboza Gutiérrez, n.d.). And, in November 2018, the FAPUV condemned the humanitarian situation of professors forced to cope with "starvation wages and very poor conditions to provide a quality education" (Aula Abierta Venezuela, 2018a, p.2; González, 2018, p.1). For many, leaving the country became a matter of their families' survival. As one of our interviewees mentioned:

> I felt that I could no longer bear to continue living in Venezuela. It was not only the economic situation or social insecurity, it was everything: food insecurity, insecurity in medicines, health, police violence, lack of basic services. It was the unworthy and inhumane treatment that the government applied to us in every way. My salary was barely enough to eat for a week, and I had to work like crazy to be able to keep myself afloat.

In 2020, with the start of the COVID-19 pandemic, the budgetary asphyxia of higher education institutions was aggravated even further (Aula Abierta Venezuela, 2020a). Public universities experienced record budgetary deficits of over 90%, as in the case of the Central University of Venezuela, or even as high as 99%, as in the case of LUZ (Aula Abierta Venezuela, 2020e; Efecto Cocuyo, 2018; Luque, 2020). This was accompanied by systematic infringements on academic freedom and university autonomy through the practices of criminalization of protest, discrimination and retaliation against university students and faculty who challenged the government's narratives, and censorship of critical debates both on university campuses and beyond (Derechos Universitarios, 2020; Gómez Gamboa et al., 2019, pp. 101–206).

In May 2020, the President of the National Constituent Assembly threatened members of the Academy of Physics, Mathematics and Natural Sciences, who issued an academic report warning about a possible increase in the number of COVID-19 cases. He requested activating Operation Tun-Tun, meaning that repressive revolutionary forces should target and intimidate academicians. Similarly, the university professor and director of the Graduate School of Medicine at LUZ, Professor Freddy Pachano, was threatened by the governor of Zulia State after speaking out on two suspected COVID-19 cases in Zulia State. At the time, the Venezuelan government was still denying the presence of the coronavirus in the country. These practices of intimidation restricted Pachano from freely speaking out on this important topic and forced him to leave the country.[6]

These are examples of how earlier patterns of the state's infringements on critical voices in academia were repeated. In 2017, Professor Santiago Guevara was arbitrarily detained and brought before a military court after his publications on the economic situation in the country. Gómez et al. (2020b) stated that in his trial, the prosecutor presented the scholar's books and academic papers as evidence of his crimes. In 2019, a resolution was issued by the National Council of Universities, which requested the initiation of a criminal inquiry against the members of the Universities Rectors Association for not recognizing the "de-facto government" of Nicolás Maduro (Aula Abierta Venezuela, 2019b; Gazeta Oficial de la República Bolivariana de Venezuela, 2019).

The decline in academic freedom under President Maduro, who continued along the line of the Chávez government and further curtailed the autonomy of universities, was also evident in the interviews we conducted for this study. A Venezuelan scholar now living in Colombia stated:

> There was an indirect dismissal from the National School of the Magistracy, by taking away the position I had. This happened without prior notice and without any authority. I'm sure this was done for being clearly anti-Chávez. This made it difficult for me to get a job in my area . . . . The situation was becoming more and more difficult . . . Those who know me think that if I had stayed in Venezuela, I would probably be in jail by now. (Personal interview, August 24, 2020)

Moreover, scholars now living in exile mentioned they had faced threats of physical violence and were forced to leave Venezuela because of fear for their safety and that of their family members:

> I received death threats aimed towards my daughters. Also, I had threatening graffiti on the door of my residence. . . . I was threatened several

---

6   After having published on his Twitter account the existence of two suspected cases of coronavirus in the Autonomous Service "Hospital Universitario de Maracaibo" (SAHUM in Spanish), Governor Prieto announced to the press that Professor Pachano had to inform the General Directorate of Military Counterintelligence (DGCIM) about his statements and said that he would immediately request the Public Ministry to open a criminal investigation against him because his statements were related to state security issues (Aula Abierta Venezuela, 2020b).

times by criminals and uniformed personnel in the service of the regime. (Personal interview, August 28, 2020)

The situation as here described has been documented in international reports on the state of academic freedom in Venezuela. The SAR's 2020 Free to Think Report (SAR, 2020, p. 111) stated that according to the Academic Freedom Index (AFI), a global analysis of national levels of respect for academic freedom, Venezuela received a score of 0.28 out of 1.00 (a "D" ranking). This was near the bottom quintile of the 140 countries evaluated and was comparable with those of countries including Libya (0.24), Rwanda (0.22), Cuba (0.14), and Yemen (0.14) and well below the average of countries in Latin America and the Caribbean (0.77) (V-Dem Institute, 2020). Thus, the major infringements on academic freedom, fear of prosecution, and security concerns have driven from the country even those academics who were not considering living abroad as a viable alternative:

> I was really one of those who refused to leave the country. I wanted to continue to have hope and optimism. However, in 2017, we were victims of a lot of robberies at the university. I gave my professional life to academic work, and seeing the university in those conditions depressed me a lot. All my colleagues told me not to continue going to the university center for security reasons. (Personal interview, August 31, 2020)

Pull factors for Venezuelan scholars in host countries

The dataset of 230 Venezuelan scholars, which we compiled for this study, allowed us to identify the top three countries that were chosen as options for living and working abroad. These were Colombia, Ecuador, and Chile. In our sample of interviewed scholars, 42.5% migrated to Colombia, 35% to Ecuador, and 22.5% to Chile. The reasons for the choice of destination ranged from geographical proximity, sociocultural and family-related factors to welcome culture in host institutions, appreciation of academic credentials, the economic situation in the country of choice, and of course prior personal connections to the receiving institutions.

Colombia was the main destination country for Venezuelan scholars who were interviewed. Of the total number of scholars consulted who moved to Colombia, 31% found employment within universities in the capital, Bogotá, and 69% within universities in other regions of Colombia. Notably, the majority of Venezuelan professors coming from regions that are close to the border with Colombia migrated to regions in Colombia that are geographically close to their region of origin and began working at Colombian universities there. Also, it is easier for displaced Venezuelans to migrate to Colombia as they have an "open-arms" approach by issuing special permits providing a protected status, work permits, and legal residency. While there was a lapse in official entries due to COVID-19, President Ivan Duque made "a welcoming border" one of his top priorities during his presidency and in April 2021 gave one million more undocumented migrants rights to legal employment, health care, education, and Colombian banking services for 10 years (Frydenlund et al., 2021). Thus, key factors of migrating to Colombia are the proximity and the easier legal pathways of entering the country.

In relation to the interviewed professors who moved to Ecuador, 54.5% started working in universities in Quito and 45.5% in other regions. Regarding professors who migrated to Chile, 80% are in Santiago and 20% in other regions. Most of the Venezuelan scholars who were interviewed stated that instead of the proximity and legal pathways as was indicated by the Venezuelan scholars who migrated to Colombia, the decision to move to Ecuador or Chile lay within prior institutional, professional, or personal connections with actors linked to the host university.

Besides the personal motivations rooted in prior connections of Venezuelan scholars to host institutions, host countries' high demand for university professionals must be emphasized as an additional pull factor. This demand grew over time with the development of education sectors in the countries concerned throughout the second half of the twentieth century. In the 1960s and 1970s, there was a strong development of higher education in most of the countries in Latin America. Expansion was funneled through the increase in the number of public universities and the multiplication of their enrollments. From 1991 to 2012, the gross enrollment ratio for the entire region increased from 17% to 43%. Chile and Colombia provide two examples: In Colombia, tertiary education enrollment grew from 14% in 1991 to 45% in 2012; in Chile, the gross enrollment rate increased even more, from 21% in 1991 to 71% in 2012 (González-Velosa et al., 2015).

Over the years, the expansion of higher education received more and more criticism by the population as a growing concern developed that this expansion in coverage would be accompanied by the massification and deterioration in quality impacting the universities. Critics have called it "academic capitalism" where there is a "capitalization of knowledge" consisting of pressures for self-funding via tuition charges, privatization of universities, and the demand on researchers and teachers to collaborate more closely with private firms (Bernasconi, 2008, p.11). In addition, the growing numbers of students enrolled in universities were not necessarily reflected in the numbers of highly qualified faculty. In fact, Colombian universities, for instance, often face shortages of teaching staff with postgraduate degrees. A Venezuelan professor working now in Colombia described the situation at his host university as follows:

> [M]ost of its professionals have only master's degrees or even pre-degrees. There are barely staff with doctorates. As you can only have a fixed contract in the Colombian university once you have a doctoral degree, I consider this a big opportunity for Venezuelan scholars. The demand here is higher, and salaries and wages are much better. (Personal interview, August 31, 2020)

Another scholar who migrated to Chile explained the demand for Venezuelan scholars as follows:

> Chile is a country that has very expensive education fees, and for this reason, there are few Chilean academics with doctoral degrees. This allows and facilitates the reception of foreign professionals. As Venezuela has many skilled doctorates, we are a highly desirable group. (Personal interview, September 6, 2020)

Another pull factor for scholars to migrate was noted by the interviewees as the changing research policies of the host countries. In many countries in South America, for a long time universities focused more on teaching and less on their research productivity. As the world has moved increasingly toward a global knowledge economy over the past years, higher education and research has taken a more prominent place on the governments' agendas. Therefore, universities are increasingly called upon to guide socioeconomic development and help solve problems in countries (van Hoof, 2014). In Ecuador, for example, the government attempted to increase the research culture among its institutions of higher education in order to solve pressing socioeconomic problems. To do so, Ecuador's Secretariat of Higher Education started the Prometeo-Viejos Sabios grant program in 2011 that continued until 2017. The program aimed to attract international scholars to the country. The offered stipends were considered generous and sufficient to attract scholars in the global marketplace. This attracted the attention of many highly qualified

Venezuelan scholars, who were the largest group of beneficiaries with 156 laureates, followed by Spain with 269 laureates (Celi, 2017).

When gaining accreditation, universities are expected to compete for research funding and to engage in doctoral education in Chile, Colombia, and Ecuador, for which having highly qualified faculty is essential. Hence, having a PhD, as many Venezuelan scholars have, gave them a great advantage to be hired over many nationals who had fewer opportunities to obtain a PhD in the past. In this regard, one of the professors interviewed stated:

> When I was hired, I thought it was because of my resume. But then I realized that the university was about to have an accreditation process and needed to have more doctors in the faculty. Since I arrived, they have only wanted to hire doctors. Those who only have a master's degree are left aside. (Personal interview, September 6, 2020)

This situation presented a good opportunity for universities in the host countries to hire highly qualified academics, but it also created economic opportunities for Venezuelan academics. Several of the interviewed scholars stated that in the universities they work at abroad, they are expected to produce publications periodically, as these universities have funds to support publication activities. They also claimed to be receiving wages that allow them to support their families, which in Venezuela was impossible (personal interviews, August 2020).

Finally, the higher level of academic freedom in the host countries and the possibility of escaping the repressive environment at home were described by the scholars as additional motives to migrate. The AFI has demonstrated the differences in the situation of academic freedom between Venezuela and the top three countries of destination. On a scale from low to high (0–1), including indicators such as freedom to research and teach, freedom of academic exchange and dissemination, institutional autonomy, campus integrity, freedom of academic and cultural expression, the situation of academic freedom in Venezuela reached 0.28, Colombia 0.64, Ecuador 0.79, and Chile 0.93 in 2019. Therefore, the higher degree of academic freedom in combination with the previously mentioned opportunities of obtaining more security, a higher salary, and the demand for faculty members with a PhD in the host countries have been the main pull factors for Venezuelan scholars.

Challenging experiences in exile

Even though the universities in Ecuador, Colombia, and Chile offer opportunities for Venezuelan professors and researchers to continue their academic work, there remain many challenges to overcome. Many of the interviewed professors mentioned difficulties with documentation issues, as it is nearly impossible to obtain official documents such as passports in Venezuela due to the high prices of renewing or obtaining them. This complicates regular migration to host countries, as noted by one interviewee:

> I have not noticed any discrimination for being Venezuelan. The limitations that currently affect me when applying to other universities is the difficulty of procedures in Venezuela. (Personal interview, September 1, 2020)

In Chile, many more barriers are placed along its legal pathway, caused by growing anti-immigrant sentiments in the country (The Economist, 2018). Moreover, a new visa option was introduced: the *Visa de Responsabilidad Democrática*. This visa provides Venezuelans with a legal pathway to permanent residency, as it allows them to reside in Chile for a year so the visa can be renewed. Acquiring this visa is very difficult, however, since the visa needs to be requested in Venezuela, and none are granted in Chile or in exterior consulates. In addition, the application requires a valid passport. This document is incredibly difficult to obtain since the current cost of renewing a passport in Venezuela can be hundreds or even thousands of dollars, depending on the level of inflation that day and what bribe is being arbitrarily asked for by the "official" in charge of renewing passports. Thus, the difficulties of obtaining a passport through official channels in Venezuela facilitate the proliferation of corruption (Guerrero, 2018; Transparencia Venezuela, 2020).

Similar restrictions are placed on the legal immigration process by Ecuador. Ecuador has maintained paths to regularization for Venezuelans, where the country has some of the most progressive human rights, migration, and asylum laws in the region, including its 2017 Human Mobility Law, which embodies an approach to regularizing the status of refugees, asylum seekers, and migrants. However, government policies have undermined the intent of the law, preventing Venezuelans from accessing their rights in practice. These policy changes appear to be politically motivated, coming in response to surges in arrivals, shifts in public opinion, and a spike in xenophobia. New entry requirements imposed in late 2018 and early 2019 closed the border to many Venezuelans, including scholars.

Therefore, it can be a time-consuming, costly process for Venezuelan scholars to fulfill all the requirements necessary to obtain a legal status in a host country (IACHR, 2018). Adding to that, the recognition of academic titles can also be complicated. As one scholar noted:

> A lot of time and money is lost in bureaucracy with all the processes that there are here [Chile]. It is very exhausting. (Personal interview, August 24, 2020)

The interviews revealed that even after finding a job in a university in the host country, there is often still a lot of uncertainty involved that is linked to the temporary contracts that are given. A professor who is now working in Ecuador said:

> It has been difficult to stay active in the university because it is on a contract basis of two contracts a year of four or five months each. There is a lot of academic rivalry. . . . Every semester you have to enter a state of anguish whether or not they are going to hire you again. (Personal interview, September 9, 2020)

A similar statement was given by a professor living in Colombia:

> In principle, when I arrived, the situation was easier. Now, with the mass exodus, it is harder to get opportunities. In my case, I especially feel anguish when I must renew the contract. There is always the uncertainty of whether or not they will do that. (Personal interview, August 21, 2020)

However, the most competitive and insecure climate for professors and researchers was found in Chile, as one interviewed noted:

> The infrastructure and facilities of the university where I work are excellent, but the whole system is fiercely competitive. In my opinion, the university does very little to create a sense of belonging. There is a lot of academic staff turnover, many part-time professors and few full-time ones. At the slightest mistake, you are fired without an explanation. No matter how well you have done previously. I feel we are often being exploited by giving us many extra activities to do in addition to teaching and research. In fact, there is almost no time for research. (Personal interview, August 31, 2020)

Amid the broader context of growing nationalism across the region, Venezuelan scholars also mentioned an increasing level of xenophobia in the host countries. This implies, among other things, a fierce competition or hostility among colleagues or, in some cases, even difficulty in entering a university. A professor who is now in Ecuador explained:

> The PhD title has been the key that opened the doors for me in Ecuador six years ago. At present, xenophobia limits the entrance of Venezuelan professors however. It has become a double-edged sword because even though we are necessary for accreditation, it is also true that they are not very willing to pay a doctor's salary. In fact, I am paid as if I only obtained my master's degree, while I also obtained a PhD. Even so, I do not complain because there is no work in Venezuela. So, I keep silent in order to survive. (Personal interview, August 24, 2020)

Similar tensions were also mentioned by Venezuelan scholars residing in Colombia and Chile. A professor who is now in Colombia explained the sense of competition and nationalism that can be noted in the universities:

> It is very usual to notice discrimination in the academia for being a foreigner. They consider that you take away the opportunity of a native professional of the country. (Personal interview, August 30, 2020)

In summary, Venezuelan scholars experience a plethora of challenges in exile. There are many barriers that they encounter on the legal pathways to enter Chile and Ecuador (to a lesser extent Colombia). Moreover, the rivalry and rising discrimination and xenophobia in the workplace in the host countries create additional insecurities for Venezuelan scholars, amplifying their already vulnerable situation.

Conclusion: Seeking cohesion and contributing to the internationalization of knowledge

As mentioned earlier, there are various positive aspects of emigration for Venezuelan scholars, such as finding more security and the fulfillment of basic needs and economic opportunities. In addition, migration to the host countries fosters further development of their professional networks and provides venues for new collaborative research projects. Many interviewed scholars mentioned the possibility to publish again, something that was very difficult to do in Venezuela due to the restrictive circumstances and low university budgets. The exchanges through joint publications with colleagues at host universities offered many opportunities for novel knowledge production. A Venezuelan professor who now works at a university in Ecuador described this phenomenon:

> Without a doubt, I consider that the contribution of foreign academics has been very important for Ecuador, mainly the arrival of Venezuelans. For example, the production of scientific articles has increased considerably in recent years. Currently, a master's degree in mathematics is taught mainly by Venezuelan mathematicians here. (Personal interview, September 1, 2020)

Drawing on their extensive training and previously gained research and teaching expertise, Venezuelan scholars bring new theoretical and methodological approaches and teaching techniques, contributing to the diversification and internationalization of knowledge production in the host academic systems. But Venezuela could also benefit from the exile experiences of the scholars should they be able to return in the future. As was mentioned by an exiled academic:

> Our work to which we were entrusted is teaching and research. Here, in exile, we are consolidating and validating our knowledge. This experience and the links between professionals makes us stronger and enriches us as individuals. So, when we return to our country, we will take our bags and a load of new experiences that will undoubtedly improve what we have in Venezuela and what we did not know how to value before. (Personal interview, September 1, 2020)

In their responses to the question of whether the Venezuelan scholars would be willing to return to their home country when possible, nearly all answered that they would want that. Additionally, exiled scholars who were interviewed were very willing to form networks abroad, have joint research programs, and find ways to keep the universities open for Venezuelan students. This transnational dynamic shows that many opportunities have been created with the aim of keeping Venezuelan universities open and supporting academia. A professor now in exile in Ecuador mentioned this:

> In support of the Venezuelan academy, I have offered to give virtual classes in the schools of law and political science of LUZ. Recently, I also helped

three colleagues from LUZ's law school to teach and participate in research projects in universities in Ecuador. (Personal interview, August 26, 2020)

These examples show the value of academic cohesion and the need to search for further ways to overcome the disconnection between Venezuelan academic communities in exile and those back home. Notably, at-risk Venezuelan scholars have received little attention in respect to the international initiatives that support academics in exile, despite the UNHCR declaration of March 11, 2018 establishing the dimension of the Venezuelan humanitarian crisis as comparable to that of Syria (IACHR, 2020; OHCHR, 2018; UNHCR, 2020). Therefore, more support programs are needed to help exiled Venezuelan academics enrich their professional skills, promote their development and professional integration in the host countries, and expedite their reintegration in Venezuela once conditions allow it.

References

Aula Abierta Venezuela (2018a). Declarada Emergencia Humanitaria compleja en la UCV. Retrieved from http://aulaabiertavenezuela.org/index.php/20 18/10/05/declarada-emergencia-humanitaria-compleja-en-la-ucv/

Aula Abierta Venezuela (2019a). Press release on University professors in resistance against the "de facto government". Retrieved from http://aula abiertavenezuela.org/index.php/2019/12/27/profesores-universitarios-en -resistencia-frente-al-gobierno-de-facto/

Aula Abierta Venezuela (2019b). Rectores e averu bajo amenaza por cuestionar gobierno de facto. Retrieved from http://aulaabiertavenezuela.org/index .php/2019/08/02/rectores-de-averu-bajo-amenaza-por-cuestionar-gobie rno-de-facto/

Aula Abierta Venezuela (2020a). Informe preliminar: Afectaciones de la educación de calidad en las universidades públicas venezoelanas en el marco del Covid-19. Retrieved from http://aulaabiertavenezuela.org/wp-content/uploads/2020/04/AFECTACIONES-A-LA-EDUCACI%C3%93N-D E-CALIDAD-EN-LAS-UNIVERSIDADES-P%C3%9ABLICAS-VENEZOLA NAS-EN-EL-MARCO-DEL-COVID-19-1.pdf

Aula Abierta (2020b). Situación de la libertad académica, la autonomía universitaria y el derecho a la educación de calidad en Venezuela en el marco del Covid-19. Retrieved from http://aulaabiertavenezuela.org/wp-

content/uploads/2020/09/INFORME-PRELIMINAR-LIBERTAD-ACADE
MICA-56-pag.pdf

Aula Abierta Venezuela (2020c). 1 Segundo informe diagnóstico: Emergencia
de la engergía eléctrica, agua y saneamiento y gestion de residuos solidos
en Venezuela (AÑO 2019- ABRIL 2020). Retrieved from http://aulaabierta
venezuela.org/wp-content/uploads/2020/05/A.A.-SEGUNDO-INFORME
-DIAGN%C3%93STICO-EMERGENCIA-DE-LA-ENERG%C3%8DA-EL%C
3%89CTRICA-AGUA-Y-SANEAMIENTO-Y-GESTI%C3%93N-DE-RESIDU
OS-S%C3%93LIDOS-EN-VENEZUELA-A%C3%91O-2019-ABRIL-2020.pdf

Aula Abierta Venezuela (2020d). 1 Informe preliminar: Propuestas de
mejoramiento de la energía eléctrica, agua y saneamiento y gestion de
residuos sólidos en Venezuela. (AÑO 2019-ABRIL 2020). Retrieved from
http://aulaabiertavenezuela.org/wp-content/uploads/2020/05/A.A-INFO
RME-PRELIMINAR_-PROPUESTAS-DE-MEJORAMIENTO-DE-LA-ENE
RG%C3%8DA-EL%C3%89CTRICA-AGUA-Y-SANEAMIENTO-Y-GESTI%C
3%93N-DE-RESIDUOS-S%C3%93LIDOS-EN-VENEZUELA-A%C3%91O-2
019-ABRIL-2020.pdf

Aula Abierta Venezuela (2020e). Asfixia presupuestaria desata migración
forzada y reducción de investigación científica en las unidades de
polímeros de la USB y LUZ. Retrieved from http://aulaabiertavenezuela.
org/index.php/2020/10/23/asfixia-presupuestaria-desata-migracion-for
zada-y-reduccion-de-investigacion-cientifica-en-las-unidades-de-polim
eros-de-la-usb-y-luz/

Barboza Gutiérrez, O. E. (n.d.). Acuerdo para la declaración de la emergencia
humanitarian compleja de la educación. Asamblea Nacional Venezuela.
Retrieved from https://www.asambleanacionalvenezuela.org/actos/detal
le/acuerdo-para-la-declaracion-de-la-emergencia-humanitaria-complej
a-de-la-educacion-300

Bastidas, G. (2018, March 13). ACNUR: Venezolanos que migran de su país
necesitan protección internacional [Video]. *Youtube.* Retrieved from https
://www.youtube.com/watch?v=Ga2GDkAio3c

Bernasconi, A. (2008). Is there a Latin American model of the university?
Comparative Education Review, 1, 27–52. https://ejournals.bc.edu/index.ph
p/ihe/article/view/8031

Celi, E. (2017). Prometeo: el proyecto que movilizó académicos de Venezuela
y España hacia el país. *Primicias.* Retrieved from https://www.primicias.e
c/noticias/politica/becarios-proyecto-prometeo/

Comisión Interamericana de Derechos Humanos (2019, February 15). Regional: Libertad académica [Video]. *Youtube*. Retrieved from https://w ww.youtube.com/watch?v=CfOvoFHGroE&t=1888s

Conte de San Blas, G. (2018, November 18). Ciencia y tecnología en el abandono. *Tal Cual*. Retrieved from https://talcualdigital.com/ciencia-y-tecnologia-en-el-abandono-por-gioconda-cunto-de-san-blas/

Derechos Universitarios (2020). Aula Abierta denounces multiple violations against academic freedom in Latin-America. Retrieved from http://derec hosuniversitarios.org/index.php/2020/06/05/aula-abierta-denounces-m ultiple-violations-against-academic-freedom-in-latin-america/

Efecto Cocuyo (2018) Gobierno aprobó solo 0,14% del presupuesto solicitado por LUZ para 2019. Retrieved from https://efectococuyo.com/la-humani dad/gobierno-aprobo-solo-014-del-presupuesto-solicitado-por-luz-para -2019/

El Nacional (2019). Instituto de Investigación aseguró que 30% de los investigadores emigraron. Retrieved from https://www.elnacional.com/s ociedad/instituto-investigacion-aseguro-que-los-investigadores-emigra ron_280007/

Frydenlund, E., Padilla, J. J., & Palacio, K. (2021, April 14). Colombia gives nearly 1 million Venezuelan migrants legal status and right to work. *The Conversation*. Retrieved from https://theconversation.com/colombia-give s-nearly-1-million-venezuelan-migrants-legal-status-and-right-to-work -155448

Gazeta Oficial de la República Bolivariana de Venezuela (2019). Agreement No. 0082 of May 30, 2019. *Official Gazette of the Bolivarian Republic of Venezuela* No. 41684, dated July 31.

Gómez Gamboa, D., Velazco, K., Villalobos, R., & Faria, I. (2019). *Libertad Académica y Autonomía Universitaria: Una mirada desde los derechos humanos. Referencias a Venezuela (2010–2019)*. Colección Textos Universitarios, Universidad del Zulia, Maracaibo, Venezuela. Retrieved from http://dere chosuniversitarios.org/wp-content/uploads/2020/02/Libertad-acad%C3 %A9mica-y-autonom%C3%ADa-universitaria-una-mirada-desde-los-der echos-humanos-Referencias-a-Venezuela-2010-2019.pdf

Gómez Gamboa, D., Villalobos, R., Velazco, K., & Van Dijk, L. (2020a). *Academic Freedom in Latin America. A Human Rights Approach: from Theory to Practice*. Aula Abierta Venezuela.

Gómez Gamboa, D., Velazco, K., & Ortega, D. (2020b). *Derecho a la libertad académica en Latinoamérica (Volumen I)*. Instituto de Filosofía del

Derecho "Dr. José M. Delgado Ocando". Facultad de Ciencias Jurídicas y Políticas de la Universidad del Zulia. Maracaibo, Venezuela. Retrieved from http://derechosuniversitarios.org/wp-content/uploads/2021/02/Lib ro_Libertad_Academica_LATAM_2020-.pdf

González, Y. (2018, November 16). Fapuv declara "Emergencia Humanitaria Educativa". *El Universal*. Retrieved from https://www.eluniversal.com/pol itica/25947/fapuv-declara-emergencia-humanitaria-educativa

González-Velosa, C., Rucci, G., Sarzosa, M., & Urzúa, S. (2015). Returns to Higher Education in Chile and Colombia. IDB Working Paper Series, No. 587. Inter-American Development Bank (IDB), Washington, DC. Retrieved from http://hdl.handle.net/11319/6858

Guerrero, Y. (2018, May 8). Los desafíos de sacar un pasaporte en Venezuela. *Prodavinci*. Retrieved from https://prodavinci.com/los-desafios-de-sacar-un-pasaporte-en-venezuela/

Hanke, S. (2019, November 13). Venezuela's Hyperinflation Drags on for a near Record- 36 Months. *Forbes*. Retrieved from https://www.forbes.com /sites/stevehanke/2019/11/13/venezuelas-hyperinflation-drags-on-for-a-n ear-record36-months/?sh=3252c3256b7b

Inter-American Commission on Human Rights – IACHR (2017a). Democratic Institutions, the Rule of Law and Human Rights in Venezuela. Retrieved from https://www.oas.org/en/iachr/reports/pdfs/Venezuela2018-en.pdf

Inter-American Commission on Human Rights – IACHR (2017b). IACHR Condemns Supreme Court Rulings and the Alteration of the Constitutional and Democratic Order in Venezuela. Retrieved from https ://www.oas.org/en/iachr/media_center/PReleases/2017/041.asp

Inter-American Commission on Human Rights – IACHR (2018). Resolution 2/18 Forced Migration of Venezuelans. Retrieved from http://www.oas.or g/en/iachr/decisions/pdf/Resolution-2-18-en.pdf

Inter-American Commission on Human Rights – IACHR (2019). Annual Report of the Special Rapporteurship on Economic, Social, Cultural And Environmental Rights (SRESCER). Retrieved from http://www.oas.org/e n/iachr/docs/annual/2019/docs/IA2019REDESCA-en.pdf

Inter-American Commission on Human Rights – IACHR (2020). Report about the in loco visit to Venezuela 2020. Retrieved from http://www.oas.org/e s/cidh/prensa/comunicados/2020/106.asp

Luque, H. (2020, October 16). Presupuesto aprobado a la ucv corresponde al 227 del monto solicitado. *UCV Noticias*. Retrieved from https://ucvnoticia

s.wordpress.com/2020/10/16/presupuesto-aprobado-a-la-ucv-correspon de-al-227-del-monto-solicitado/

Office of the United High Commissioner for Human Rights – OHCHR (2018). Report: Human Rights Violations in the Bolivarian Republic of Venezuela: A downward spiral with no end in sight. Retrieved from https://www.ohc hr.org/Documents/Countries/VE/VenezuelaReport2018_EN.pdf

Organization of American States – OAS (2018). OAS Resolution D-032/18. Retrieved from https://www.oas.org/es/centro_noticias/comunicado_pre nsa.asp?sCodigo=D-032/18

Programa Venezolano de Educación Acción en Derechos Humanos (2017). Derecho a la Educación. Retrieved from https://provea.org/wp-content/ uploads/06Educaci%C3%B3n-1-2.pdf

QS (2020a). QS Latin American University Rankings. Retrieved from https:// www.topuniversities.com/university-rankings/latin-american-universit y-rankings/2020

QS (2020b). QS World University Rankings. Retrieved from https://www.top universities.com/university-rankings/world-university-rankings/2020

Red Iberoamericana de indicadores de ciencia y tecnología (2018). Publicaciones Venezuela 2009-2018. Retrieved from http://app.ricyt. org/ui/v3/bycountry.html?country=VE&subfamily=CTI_BIB&start_year= 2009&end_year=2018

Requena, J., & Caputo, C. (2016). Pérdida de talento en Venezuela: migración de sus investigadores. *Interciencia*, 41(7), 444–453. https://www.redalyc.or g/comocitar.oa?id=33946267002

Scholars at Risk – SAR (2020). Free to Think. Report of the Scholars at Risk Academic Freedom Monitoring Project. Retrieved from https://www.sch olarsatrisk.org/wp-content/uploads/2020/11/Scholars-at-Risk-Free-to-T hink-2020.pdf

Scimago Institutions Rankings (n.d.-a). Country Rankings. Retrieved from h ttps://www.scimagojr.com/countryrank.php?region=Latin%20America

Scimago Institutions Rankings (n.d.-b). Higher Education- Rankings Latin America. Retrieved from https://www.scimagoir.com/rankings.php?sect or=Higher%20educ.&country=Latin%20America&year=2004

The Economist (2018). Chile gives immigrants a wary welcome. Retrieved from https://www.economist.com/the-americas/2018/04/12/chile-gives-i mmigrants-a-wary-welcome

Transparencia Venezuela (2020). Avanza denuncia por corrupción y extorsión en el SAIME. Retrieved from https://transparencia.org.ve/avanza-denuncia-por-corrupcion-y-extorsion-en-el-saime/

United Nations Human Rights Council – UNHCR (2020). Report of the Independent International Fact-Finding Mission on the Bolivarian Republic of Venezuela. Retrieved from https://www.ohchr.org/Documents/HRBodies/HRCouncil/FFMV/A_HRC_45_33_AUV.pdf

University of Los Andes Human Rights Observatory (2020). Reporte Mensual: Situación de las Universidades en Venezuela. Retrieved from https://www.uladdhh.org.ve/wp-content/uploads/2020/06/Reporte-mensual-Situaci%C3%B3n-UniVE-MAY2020.pdf

Van Hoof, H. (2014). Ecuador's Efforts to Raise Its Research Profile: The Prometeo Program Case Study. *Journal of Hispanic Higher Education*, 14(1), 56-68. doi:10.1177/1538192714543664

V-Dem Institute (2020). Database available on: https://www.v-dem.net/en/analysis/MapGraph/

VenEuropa Canal (2018, March 7). DE VENEZUELA. ONU Alto Comisionado para Derechos Humano [Video]. *Youtube*. Retrieved from https://www.youtube.com/watch?v=N4wdKwjI75s

Worldometers (n.d.). World population- Latin America and the Caribbean Population 2021. Retrieved from https://www.worldometers.info/world-population/latin-america-and-the-caribbean-population/

List of contributors

MSc Lizzy Anjel-van Dijk holds a Master degree in International Development Studies from Utrecht University, the Netherlands. Currently, she works at Peace Brigades International where she is a program officer of Shelter City Utrecht. This programme is a global movement that offers safe and inspiring spaces to human rights defenders at risk where they can re-energise, exchange experiences, spread awareness, and expand their network with new allies. Next to that, Lizzy has a keen interest in designing and implementing knowledge sharing and advocacy strategies. Her research comprise migration topics within the Latin American context and that of academic freedom.

Dr. Vera Axyonova is a Marie Skłodowska Curie REWIRE Fellow at the University of Vienna and Principal Investigator of the project "Expert Knowledge in Times of Crisis – Uncovering Interaction Effects between Think Tanks, Media and Politics beyond Liberal Democracies". Previously, Vera worked in research, science management and policy consulting, including as Managing Director of "Academics in Solidarity", a transnational mentoring program for at-risk scholars at Freie Universität Berlin, Assistant Professor for International Integration at Justus Liebig University Giessen, and Hurford Next Generation Fellow with the Carnegie Endowment's Euro-Atlantic Security Initiative.

Dr. Nahed Ghazzoul is assistant professor of Linguistics. Currently, she is a full-time researcher at Paris Nanterre University in the MoDyCo Research Centre – PAUSE Programme She holds a PhD in Linguistic from Lancaster University, UK from 2008; a Certificate in Learning and Teaching in Higher Education, from Lancaster University, UK from 2006; an MA in Linguistics (TESOL), University of Surrey, UK from 2004. She worked at different educational places including Aleppo University/Syria, Lancaster

University/UK, and Jerash and Alzaytoonah Universities/Jordan. She was a post-doctorate-researcher at Columbia University, Columbia Global Centre/ Amman. Nahed is also a Syrian activist and human rights defender. She advocates for the rights of the refugees in getting tertiary education. She also calls for the rights of displaced Syrian academics in teaching, and research to keep their educational and vocational life running.

Dr. David Gómez Gamboa is an associate professor at the School of Law of the University of Zulia (Venezuela). He has a degree in Law from the University of Zulia, 2000, and in Journalism from the Catholic University Cecilio Acosta, 2007. He did postgraduate studies in Human Rights at the University Complutense of Madrid and in Political Science at the University of Zulia. Currently, he is the founding director of Aula Abierta, a non-governmental organization that promotes academic freedom and university autonomy in Latin-America, and coordinator of the Law School's Human Rights Commission at the University of Zulia. Also, David is a human rights activist in the context of academic freedom and university autonomy. With his NGO, Aula Abierta, he denounces human rights violations in the Latin American university context.

Dr. Olga Hünler received her PhD in clinical psychology from Middle East Technical University in 2007. Between 2016 and 2019, she was granted the Philipp Schwartz Fellowship founded by the Alexander von Humboldt Foundation, and she was hosted at Bremen University, Germany. Between 2019 and 2021 she was a fellow of the Academy in Exile's Critical Thinking Residency Program at Freie Universität Berlin, Germany. She was also an international guest researcher at the Margherita von Brentano Center for Gender Studies. She is currently an associate professor at Acibadem University, İstanbul.

Dr. Florian Kohstall is founder of "Academics in Solidarity" and head of the "Global Responsibility" Unit of Freie Universität Berlin. His research focuses on varieties of internationalization and the politics of higher education reform in the Middle East and North Africa. He taught political science in Aix-en-Provence, Cairo and Lyon and is a former research fellow of CEDEJ in Cairo and alumnus of AGYA – The Arab German Young Academy of Sciences and Humanities.

Dr. Maggi W.H. Leung is professor in International Development Studies at the Department of Geography, Planning and International Development Studies of the University of Amsterdam. Her research focuses on the uneven geographies of migration, mobilities and development, internationalization of education and labor regimes, knowledge and skill (im)mobilities, Chinese transnationalism, investments and engagements for/by/with newcomers in "shrinking" regions in Europe as well activism against racialised injustice. She has published on these topics in a range of geography and social science journals. She is one of the editors of *Geoforum* and has served as a guest editor for special issues on topics regarding migration and skills, Chinese transnationalism and welcoming spaces in Europe.

Dr. Isabella Löhr is a historian of Modern Europe in global perspective. She works on international student mobility, on knowledge production about exile and on the history of global connections and international law. Currently, she is deputy director of the French-German Centre Marc Bloch Berlin, an interdisciplinary research centre for social sciences that is dedicated to research of Europe.

Dr. Ergün Özgür is an Einstein Guest Researcher at Freie Universität Berlin. She was a research fellow at the Leibniz-Zentrum Moderner Orient, The Käte Hamburger Center for Advanced Study in the Humanities "Law as Culture", visiting professor at the Université Catholique de Louvain, Belgium, and assistant professor at Cyprus International University, Nicosia. She holds a PhD from Marmara University İstanbul, Organizational Behaviour Department. Her research interest centers on culture, ethnicity, identity, gender, migration, refugees, values, the Caucasus, the Abkhazians and the Circassians.

Dr. Carola Richter is professor for international communication at Freie Universität Berlin, Germany. In her research, she focuses on media systems and communication cultures in the Middle East and North Africa region, media and migration, foreign news coverage as well as on public diplomacy. She is director of the Center for Media and Information Literacy (CeMIL) at Freie Universität Berlin. Her latest publication is the open access book *Arab Media Systems* (2021, co-edited with Claudia Kozman, Open Book Publishers). She is involved in the initiative Academics in Solidarity at Freie Universität Berlin, supporting scholars at-risk.

Dr. Azade Seyhan holds a PhD in German Philosophical Traditions from the University of Washington, Seattle. She is Research Professor in German Studies at Bryn Mawr College in Pennsylvania. Her most recent book is *Heinrich Heine and the World Literary Map: Redressing the Canon* (Palgrave Macmillan).

Dr. Asli Telli is associate professor and a scholar of media anthropology and social informatics. She taught and advised international graduate students in numerous universities in Turkey, Switzerland, Malta, France, Germany and the US. She has published widely on participatory cultures, digital knowledge commons and dissent movements. Asli is an active member of Academics for Peace Germany and Off-University, both grassroots organizations, founded by exiled researchers. She is also a former Research Associate at Locating Media Program of University of Siegen, Germany and is currently transitioning to her new remote affiliation with WISER of Wits University, South Africa.

Dr. Rika Theo is a researcher, information specialist, and trained archivist. Previously worked as a journalist in Indonesia, she has written on a wide range of topics on economy, politics and social issues. She completed her MA in International Political and Economy and Development from the International Institute of Social Sciences The Hague, MA in Archival and Information Studies from the University of Amsterdam and her PhD in international mobility and development from Utrecht University. Her research interests comprise decolonization and social justice in the archives, knowledge mobilities and activism, as well as China-Indonesia relations and transnationalism. Currently she works as an Information Specialist for Political Science at the University of Amsterdam and dedicates her research on the inclusivity and accessibility of the Indonesian displaced archives in the Netherlands.

Dr. Aslı Vatansever holds a PhD from the University of Hamburg from 2010. She is a sociologist of work and social stratification with a focus on precarious academic labor. Her ongoing research project at *Bard College Berlin* investigates forms of academic labor activism in Europe. Her books include *Ursprünge des Islamismus im Osmanischen Reich. Eine weltsystemanalytische Perspektive* (Hamburg: Dr. Kovač, 2010), *Ne Ders Olsa Veririz. Akademisyenin Vasıfsız İşçiye Dönüşümü (Ready to Teach Anything. The Transformation of the Academic into Unskilled Worker*, Istanbul: İletişim, 2015 – co-authored with Meral Gezici-

Yalçın), and *At the Margins of Academia. Exile, Precariousness, and Subjectivity*
(Brill, 2020).

Social Sciences

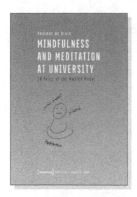

kollektiv orangotango+ (ed.)
This Is Not an Atlas
A Global Collection of Counter-Cartographies

2018, 352 p., hardcover, col. ill.
34,99 € (DE), 978-3-8376-4519-4
E-Book: free available, ISBN 978-3-8394-4519-8

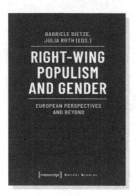

Gabriele Dietze, Julia Roth (eds.)
Right-Wing Populism and Gender
European Perspectives and Beyond

April 2020, 286 p., pb., ill.
35,00 € (DE), 978-3-8376-4980-2
E-Book: 34,99 € (DE), ISBN 978-3-8394-4980-6

Mozilla Foundation
Internet Health Report 2019

2019, 118 p., pb., ill.
19,99 € (DE), 978-3-8376-4946-8
E-Book: free available, ISBN 978-3-8394-4946-2

**All print, e-book and open access versions of the titles in our list
are available in our online shop www.transcript-publishing.com**

Social Sciences

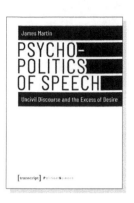

James Martin
Psychopolitics of Speech
Uncivil Discourse and the Excess of Desire

2019, 186 p., hardcover
79,99 € (DE), 978-3-8376-3919-3
E-Book:
PDF: 79,99 € (DE), ISBN 978-3-8394-3919-7

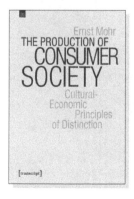

Michael Bray
Powers of the Mind
Mental and Manual Labor
in the Contemporary Political Crisis

2019, 208 p., hardcover
99,99 € (DE), 978-3-8376-4147-9
E-Book:
PDF: 99,99 € (DE), ISBN 978-3-8394-4147-3

Ernst Mohr
The Production of Consumer Society
Cultural-Economic Principles of Distinction

April 2021, 340 p., pb., ill.
39,00 € (DE), 978-3-8376-5703-6
E-Book: available as free open access publication
PDF: ISBN 978-3-8394-5703-0

**All print, e-book and open access versions of the titles in our list
are available in our online shop www.transcript-publishing.com**

Printed in the USA
CPSIA information can be obtained
at www.ICGtesting.com
JSHW011520221024
72172JS00015B/126